Jutta Arrenberg

AF220325

Analysis of Multivariate Data with SPSS

Workbook with Detailed Examples

1st Edition

Analysis of Multivariate Data with SPSS

Workbook with Detailed Examples

of

Prof. Dr. Jutta Arrenberg

1st edition

Bibliografische Information der Deutschen Nationalbibliothek

Die Deutsche Nationalbibliothek verzeichnet diese Publikation in der Deutschen Nationalbibliografie; detaillierte bibliografische Daten sind im Internet über <http:\\ dnb.dnb.de> abrufbar.

Cover-Grafik: Burkhard Arrenberg, Hamburg
Herstellung und Verlag: BoD - Books on Demand, Norderstedt
ISBN: 978 - 3 - 751 - 98971 - 8

Preface

The book presents the most important statistical inferences for multivariate data sets:

> Binary Logistic Regression, Cluster Analysis, Jarque-Bera Test, Kruskal-Wallis Test, Levene Test, Lilliefors Test, Linear Regression Analysis, Measures of Association, Multinomial Logistic Regression Analysis, Multiple Linear Regression Analysis, Ordinal Regression Analysis, Pearson Chi-Square Test, Principal Components Analysis, Shapiro-Wilk Test, t-Test, Welch Test.

Every statistical inference is explained using an easy-to-understand example to ensure learning success. The steps for finding solutions with SPSS are shown in the book. For a deeper understanding, some of the statistical methods are also explained in detail.

After reading the book, the reader is able to perform comprehensive statistical analyzes of multivariate data sets.

Previous knowledge of statistics is helpful for reading the book, but is not essential. The presented methods are plausible even without prior knowledge of statistics.

This book was written over a period of twenty semesters and benefited from the questions posed by the listeners to my lectures.

I would like to thank everyone who contributed to the development of this book and I hope you enjoy reading it.

Cologne, September 2020 Jutta Arrenberg

Contents

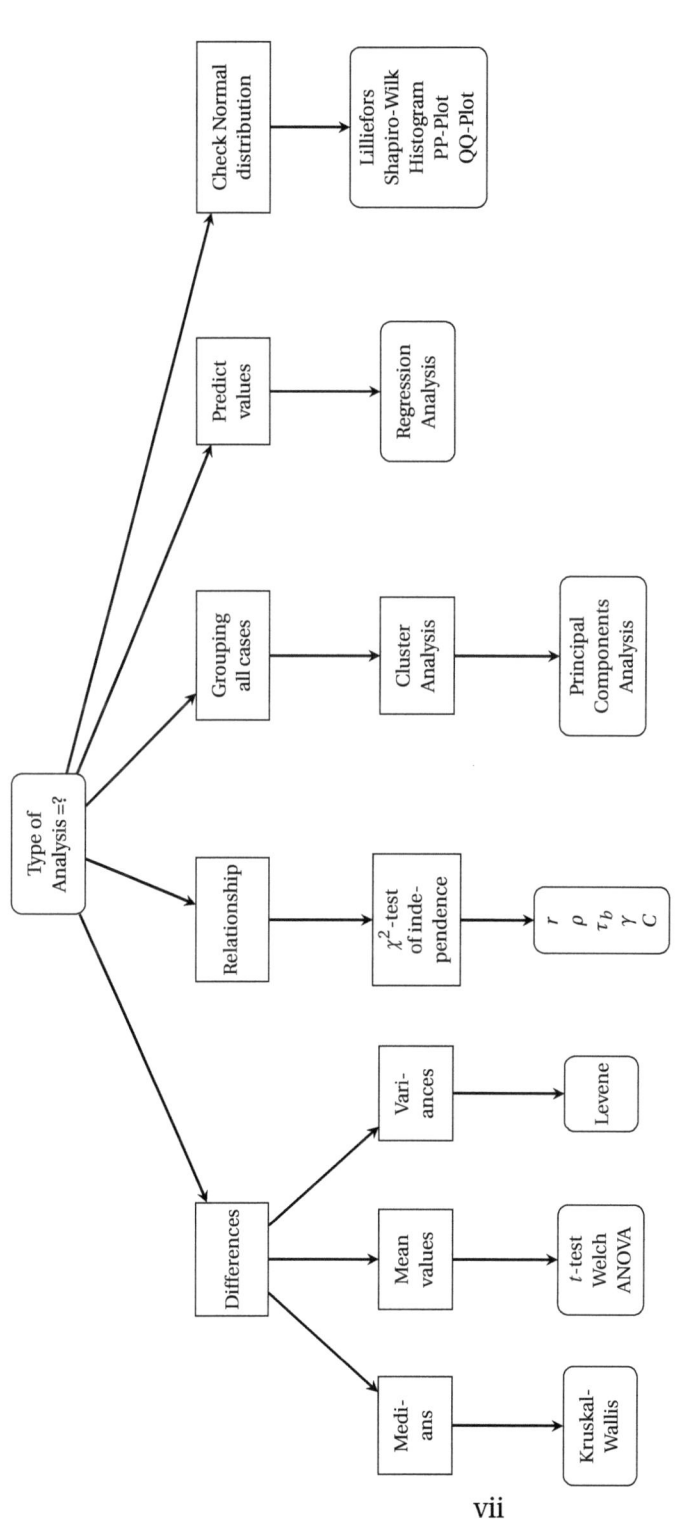

viii

1 Introduction

For a data analysis, both statistics and computer science skills are required. Computer science methods are needed to prepare, clean, convert and model data.

In this book we want to examine data sets by using the software package SPSS Statistics 25.0 (formerly: Predictive Analysis SoftWare) that provides a powerful statistical analysis.

⚠ Pressing the keys < Ctr > + < · > simultaneously you can interrupt a running process of SPSS. This is important, if you have selected wrong commands in SPSS and you are waiting and waiting for the output of SPSS. You can select the language of SPSS as follows:

1) Edit → Options

2) Language
   Output = English
   User Interface = English

3) ok

Please ask the information technology service center of your university for a free license of SPSS to write your thesis.

1.1 Levels of Variables

We have many statistical methods to analyze data sets. Most of all the methods depend on the level of the considered variables.

Variables are measured at nominal, ordinal or scale levels.

A variable can be treated as **nominal** when its values represent categories with no intrinsic ranking; for example, the department of the company where an employee works. Examples of nominal variables include X = Religious affiliation (Christian, Buddhist, Moslim ...), X = Eye color (blue, brown, green), X = Country of birth (Germany, Cameroon, Turkey ...).

A special case of a nominal variable is a **dichotomous** variable, a term for a variable that has two possible values; for example, the responses yes and no, or X = Gender (female, not female). A dichotomous variable is called a **binary** variable if the two possible outcomes of the variable are denoted by the values 0 and 1.

A variable can be treated as **ordinal** when its values represent categories with some intrinsic ranking. However, the distance between two realizations cannot be measured. For example X= Degree of service satisfaction (from highly dissatisfied to highly satisfied), X = Level of importance (from very important to unimportant). Or X= Smoking habit (non-smoker, moderate smoker, heavy smoker). Examples of ordinal variables include attitude scores representing degree of satisfaction or confidence and preference rating scores.

A variable can be treated as **scale** when its values represent ordered categories with a meaningful metric, so that distance comparisons between values are appropriate. Examples of scale variables include X = Age in years X = Income in thousands of dollars, X = Temperature.

1.2   Types of Variables

We discriminate between two types of statistical variables: discrete variables and continuous variables.

A variable X will be defined to be **discrete** if X takes on a finite or denumerable set of values.

Example 1.1
The variable X="Age of a person measured in rounded years" takes the values 0,1,2,3, ... 120. These 121 different values are a finite set of values, and X is a discrete variable.

The variable X="Gender" takes the values 1=female, 2=male, 3=third. That

are three different values, and X is a discrete variable.
The variable X="Number of workers of a company" takes the values 1,2,3,
$\ldots=\mathbb{N}$, the values of $\mathbb{N}$ are countable, so X is a discrete variable.

In contrast a variable X will be defined to be **continuous** if X takes on a nondenumerable set of values.

Example 1.2
The variable X="Age of a person" takes the values in the interval $[0;120]$. The values of this interval are not denumerable, and X is a continuous variable.
The variable X="Face Time measured in hours per day" takes values in the interval $[0;24]$ and X is a continuous variable.

Problem: We are only able to measure values in discrete steps.

1.3   Dimensions of a Data Set

A sample respectively a data set may have the following dimensions: univariate, bivariate or multivariate.
If we ask n corporations about the amount X of workers we get n values $x_1, x_2,\ldots, x_n$. The sample $x_1, x_2,\ldots, x_n$ is called **univariate** data set.
But if we have an additional question about the amount Y of the managers we get the sample $(x_1, y_1), (x_2, y_2), \ldots, (x_n, y_n)$. The sample (x_1, y_1), (x_2, y_2) $\ldots, (x_n, y_n)$ is called **bivariate** data set.
If we add a further question about the sales Z we get the sample (x_1, y_1, z_1), $(x_2, y_2, z_2), \ldots, (x_n, y_n, z_n)$. The sample $(x_1, y_1, z_1), (x_2, y_2, z_2)\ldots, (x_n, y_n, z_n)$ is called **multivariate** data set. If we have four or more questions for every of the n corporations the connecting sample is called multivariate data set, too.

1.4   Export to WORD

If you are writing a text in WORD you can export SPSS input and SPSS output without any problems to WORD. There are three options to do this.

Firstly, SPSS input and SPSS output can be exported with the commands <Ctr> + <C> and <Ctr> + <V> to WORD documents ("copy" and "paste"). Secondly, SPSS output can be exported to WORD as follows:

1) Start WORD.

2) Start SPSS and make the desired SPSS output.

3) From the left-hand box in the SPSS output window choose the output you want to export with the left mouse-pad.

4) From the menus choose:
   Edit → Copy
   A copy of the output is saved.

5) Press the two keys |Alt| and |⇆| simultaneously. Keep the key |Alt| pressed and change to WORD in that way.

6) From the WORD-Menus choose:
   Bearbeiten → Inhalte einfügen

7) From the Pulldown menus choose "Grafic". Click "OK".
   The selected output will be exported to WORD.

If it does not work, there is a third option to export a SPSS output to WORD. In this third option a new WORD document will be opened.

1) From the left-hand box in the SPSS output window choose the output you want to export.

2) On the top of the page from the menus click "Export". (Second row from above, fifth symbol)
   Under "filename" type the path of the new WORD document where you want to export the SPSS-output.
   Select as type "WORD/RTF-file(*.doc)".

3) Click "OK".
   A new WORD document with the selected output will be saved under the name you have selected in "filename".

If a SPSS table is to large for width of a page in a WORD document, the SPSS table can be fitted as follows:

1) Export the SPSS output to WORD.

2) Right mouse click on the position symbol ⊞.

3) Autoanpassen → Größe an Fenster anpassen

1.5 Summary

We discriminate between two types of variables: discrete and continuous and several levels:

Level	Type	Variable
nominal	discrete	X=Gender 1=female 2=male 3=third
ordinal	discrete	X=Degree of satisfaction 1=very satisfied 2=satisfied 3=not satisfied
scale	discrete	X=Number of siblings X=0,1,2,3, …
	continuous	X=Income (in €) $X \in [0; \infty)$

2 Statistical Analysis

Statistical analysis is split into two sections: descriptive statistics and inductive statistics.

In the case of descriptive statistics we want to compute empirical key data of a sample. Empirical key data are for example the average value or the empirical standard deviation etc. Empirical key data give a value in summary of a sample.

Example 2.1

If we have asked five people about their ages and we have got the following values: 29, 31, 25, 26, 29.

The **average** age is:

$$\frac{1}{5}[29 + 31 + 25 + 26 + 29] = 28 \text{ years}$$

And the **standard deviation** is:

$$\sqrt{\frac{1}{5}[(29-28)^2 + (31-28)^2 + (25-28)^2 + (26-28)^2 + (29-28)^2]}$$
$$= 2.19 \text{ years}$$

In contrast to descriptive statistics the purpose of inductive statistics is to give some information not about the sample but about the population of the sample. For example we want to know the mean value age of all persons in the population. Inductive statistics deals with generalizations, predictions, estimations, and test decisions from data initially presented.

2.1 Statistical Tests

A **statistical test** is a decision rule between two opposed statements H_0 and H_1. The test decides by aid of a sample. Because the reality is unknown the test can make a wrong decision:

	Reality	
	H_0 is true	H_1 is true
Test-decision on H_0	right decision	error of the 2. kind
Test-decision on H_1	error of the 1. kind	right decision

We would like that both error-probabilities are very small, but this is impossible: if the probability of one error decreases the probability of the other error will increase. Though in statistical inferences it was arranged that only the probability of the error of the first kind is very small, i.e. the probability of the first kind error is not larger than α (*read: alpha*):

$$P(\text{error of first kind}) \leq \alpha$$

Common values of α are 0.01 resp. 0.05 resp. 0.10 (vgl. Bamberg, Statistik, Kapitel 14.2). The value of α is called the theoretical **significance** level:

significance α		
0.01	0.05	0.10

The smaller the value of α the less influences are significant.
The probability of the error of the second kind is called β (*read: beta*). This probability in not larger than $1 - \alpha$; i.e. $\beta \leq 1 - \alpha$:

$$P(\text{error of second kind}) \leq 1 - \alpha$$

As a value for α we will assume $\alpha = 0.05$.
The SPSS-Software computes by aid of the sample the so called **p-value**.
⚠ The *p*-value is denoted by SPSS as empirical significance level.

> We reject the null hypothesis H_0 if and only if the *p*-value is equal to or smaller than 0.05.

The *p*-value depends on the sample and is the smallest value of α for which we can reject the null hypothesis of a test at level α. (If the *p*-value is low, than H_0 must go.)

At the moment (cf. Süddeutsche Zeitung of 23./24.09.2017 page 37, "'Das magische P'") there is a lively discussion about the meaning of a p-value (cf. Amrhein/Greenland/McShane [2019]).

With a p-value, computed by a software, a non-statistician is able to run a statistical test. However there is a risk that the findings are only considered in the two outcomes "significant" and "not significant". To avoid this mistake please also publish the height of the p-value to document how full or how short the test decision was. The smaller the p-value, the less plausible the null hypothesis.

2.2 Summary

A statistical test at level α is a decision rule based on a sample between two statements:

> H_0: Statement
> versus
> H_1: Opposed statement
> Rejection of H_0 if and only if p-value $\leq \alpha$

The probability α is an upper bound of the probability of the error that the test rejects H_0 but in reality H_0 is true. The p-value is the smallest value of α for which we can reject the null hypothesis.

3 Test Of Stochastic Independence

Main purpose: We had to detect stochastic dependence of two variables. Managerial decisions often are based on the relationship between two or more variables. For example, after considering the relationship between advertising expenditures and sales, a marketing manager might attempt to predict sales for a given amount of advertising expenditure.

Two random variables X and Y are defined to be **stochastically independent** if and only if $P(X \leq x \cap Y \leq y) = P(X \leq x) \cdot P(Y \leq y)$.

Example 3.1

We toss the dice, twice. Let denote:

$\qquad X$ = number of the first die
$\qquad Y$ = number of the first die

Then X takes on each value of the set $\{1,2,3,4,5,6\}$ with probability $\frac{1}{6}$. For Y it is the same.

Altogether there are 36 different number combinations of the two dice. And every combination has the same chance $\frac{1}{36}$ to occur, for example $P(X = 3 \cap Y = 4) = \frac{1}{36}$. Furthermore $P(X = 3) \cdot P(Y = 4) = \frac{1}{6} \cdot \frac{1}{6} = \frac{1}{36}$. The variables X and Y are stochastically independent.

Further let denote:

$$Z = \text{maximum of the two numbers } Y \text{ and } Y$$

In order for the maximum of the two numbers to be 1, both the first and the second die must each have an number of 1.

So that the maximum of the two numbers is 2, one of the three combinations $(1;2)$ or $(2;1)$ or $(2;2)$ must be rolled.

So that the maximum of the two numbers is 3, one of the five combinations $(1;3)$ or $(2;3)$ or $(3;3)$ or $(3;2)$ or $(3;1)$ must be rolled.

So that the maximum of the two numbers is 4, one of the seven combinations $(1;4)$ or $(2;4)$ or $(3;4)$ or $(4;4)$ or $(4;3)$ or $(4;2)$ or $(4;1)$ must be rolled.

So that the maximum of the two numbers is 5, one of the nine combinations $(1;5)$ or $(2;5)$ or $(3;5)$ or $(4;5)$ or $(5;5)$ or $(5;4)$ or $(5;3)$ or $(5;2)$ or $(5;1)$ must be rolled.

So that the maximum of the two numbers is 6, one of the eleven combinations $(1;6)$ or $(2;6)$ or $(3;6)$ or $(4;6)$ or $(5;6)$ or $(6;6)$ or $(6;5)$ or $(6;4)$ or $(6;3)$ or $(6;2)$ or $(6;1)$ must be rolled.

Then Z takes on each value of the set $\{1,2,3,4,5,6\}$ with probabilities:

z	1	2	3	4	5	6
$P(Z=z)$	$\frac{1}{36}$	$\frac{3}{36}$	$\frac{5}{36}$	$\frac{7}{36}$	$\frac{9}{36}$	$\frac{11}{36}$

The probabilities $P(\{X = x\} \cap \{Z = z\})$ of the events $\{X = x, Z = z\}$ are:

$x\backslash z$	1	2	3	4	5	6
1	$\frac{1}{36}$	$\frac{1}{36}$	$\frac{1}{36}$	$\frac{1}{36}$	$\frac{1}{36}$	$\frac{1}{36}$
2	0	$\frac{2}{36}$	$\frac{1}{36}$	$\frac{1}{36}$	$\frac{1}{36}$	$\frac{1}{36}$
3	0	0	$\frac{3}{36}$	$\frac{1}{36}$	$\frac{1}{36}$	$\frac{1}{36}$
4	0	0	0	$\frac{4}{36}$	$\frac{1}{36}$	$\frac{1}{36}$
5	0	0	0	0	$\frac{5}{36}$	$\frac{1}{36}$
6	0	0	0	0	0	$\frac{6}{36}$

We see that as follows:

Event	number $(X; Y)$ of the two dice
$\{X = 1\} \cap \{Z = 1\}$	(1;1)
$\{X = 1\} \cap \{Z = 2\}$	(1;2)
$\{X = 1\} \cap \{Z = 3\}$	(1;3)
$\{X = 1\} \cap \{Z = 4\}$	(1;4)
$\{X = 1\} \cap \{Z = 5\}$	(1;5)
$\{X = 1\} \cap \{Z = 6\}$	(1;6)
$\{X = 2\} \cap \{Z = 1\}$	impossible
$\{X = 2\} \cap \{Z = 2\}$	(2;1) or (2;2)
$\{X = 2\} \cap \{Z = 3\}$	(2;3)
$\{X = 2\} \cap \{Z = 4\}$	(2;4)
$\{X = 2\} \cap \{Z = 5\}$	(2;5)
$\{X = 2\} \cap \{Z = 6\}$	(2;6)
$\{X = 3\} \cap \{Z = 1\}$	impossible
$\{X = 3\} \cap \{Z = 2\}$	impossible
$\{X = 3\} \cap \{Z = 3\}$	(3;1) or (3;2) or (3;3)
$\{X = 3\} \cap \{Z = 4\}$	(3;4)
$\{X = 3\} \cap \{Z = 5\}$	(3;5)
$\{X = 3\} \cap \{Z = 6\}$	(3;6)
$\{X = 4\} \cap \{Z = 1\}$	impossible
$\{X = 4\} \cap \{Z = 2\}$	impossible
$\{X = 4\} \cap \{Z = 3\}$	impossible
$\{X = 4\} \cap \{Z = 4\}$	(4;1) or (4;2) or (4;3) or (4;4)
$\{X = 4\} \cap \{Z = 5\}$	(4;5)
$\{X = 4\} \cap \{Z = 6\}$	(4;6)
$\{X = 5\} \cap \{Z = 1\}$	impossible
$\{X = 5\} \cap \{Z = 2\}$	impossible
$\{X = 5\} \cap \{Z = 3\}$	impossible
$\{X = 5\} \cap \{Z = 4\}$	impossible
$\{X = 5\} \cap \{Z = 5\}$	(5;1) or (5;2) or (5;3) or (5;4) or (5;5)
$\{X = 5\} \cap \{Z = 6\}$	(5;6)
$\{X = 6\} \cap \{Z = 1\}$	impossible
$\{X = 6\} \cap \{Z = 2\}$	impossible
$\{X = 6\} \cap \{Z = 3\}$	impossible
$\{X = 6\} \cap \{Z = 4\}$	impossible
$\{X = 6\} \cap \{Z = 5\}$	impossible
$\{X = 6\} \cap \{Z = 6\}$	(6;1), (6;2), (6;3), (6;4), (6;5), (6;6)

But the common probability of (X, Z) does not equal the product of the single probabilities, for example:

$$P(X = 3 \cap Z = 4) = P(X = 3 \cap Y = 4) = \frac{1}{36}$$

and:

$$P(X = 3) \cdot P(Z = 4) = \frac{1}{6} \cdot \frac{7}{36} = \frac{7}{216} \neq \frac{1}{36}$$

This means that the variables X and Z are stochastically dependent.

For many statistical inferences it is assumed that the variables are stochastically independent. The Pearson Chi-Square test of independence verifies this assumption based on an bivariate data set.

3.1 Pearson Chi-Square Test

We make the test decision based on a bivariate data set of the two random variables. The levels of the two random variables must be as follows:

nominal:	yes
ordinal:	yes
scale:	yes

In this section we discuss the common practical problem of determining if there is a dependence between a pair X, Y of random variables.

The testing problem of Pearson Chi-Square Test is:

Pearson's Chi-Square Test
H_0 : X, Y are stochastically independent
versus
H_1 : X, Y are dependent
Rejection of $H_0 \Leftrightarrow p$-value $\leq \alpha$

Example 3.2 (*Partei_Eltern.sav* Agresti [1990] p. 32)
We want to know whether the party identification X of a student depends on the party identification Y of his parents. The level $\alpha = 0.05$ testing problem is:

H_0 : Student and parents party identification are stochastically independent

H_1 : Student and parents party identification are dependent

In an opinion survey regarding party identification there was some question as to whether or not the party identification of a student might view issue differently from the party identification of his parents. We asked 1 852 high school students:

Parent Party Identification	Student Party Identification		
	Democrat	Independent	Republican
Democrat	604	245	67
Independent	130	235	76
Republican	63	180	252

Summing across all rows $n_{i\bullet}$ and summing across all columns $n_{\bullet j}$ we get:

Parent Party of Identification	Student Party Identification			Σ
	Democrat	Independent	Republican	
Democrat	604	245	67	916
Independent	130	235	76	441
Republican	63	180	252	495
Σ	797	660	395	1 852

We reject the null hypothesis if the differences between the observed frequencies 604, 245, ..., 252 and the under H_0 **expected frequencies** $\frac{916 \cdot 797}{1852}$, $\frac{916 \cdot 660}{1852}$, ..., $\frac{495 \cdot 395}{1852}$ are too "large". This means that we reject the null hypothesis H_0 if and only if the p-value is equal to or less than $\alpha = 0.05$. We get the observed value (i.e. the empirical value) of the test statistic as follows:

15

$$\chi^2_{\text{emp.}} = \sum_{i=1}^{I} \sum_{j=1}^{J} \frac{\left(n_{ij} - \frac{n_{i\bullet} \cdot n_{\bullet j}}{n}\right)^2}{\frac{n_{i\bullet} \cdot n_{\bullet j}}{n}}$$

$$= \frac{\left(604 - \frac{916 \cdot 797}{1852}\right)^2}{\frac{916 \cdot 797}{1852}} + \frac{\left(245 - \frac{916 \cdot 660}{1852}\right)^2}{\frac{916 \cdot 660}{1852}} + \frac{\left(67 - \frac{916 \cdot 395}{1852}\right)^2}{\frac{916 \cdot 395}{1852}}$$

$$+ \frac{\left(130 - \frac{441 \cdot 797}{1852}\right)^2}{\frac{441 \cdot 797}{1852}} + \frac{\left(235 - \frac{441 \cdot 660}{1852}\right)^2}{\frac{441 \cdot 660}{1852}} + \frac{\left(76 - \frac{441 \cdot 395}{1852}\right)^2}{\frac{441 \cdot 395}{1852}}$$

$$+ \frac{\left(63 - \frac{495 \cdot 797}{1852}\right)^2}{\frac{495 \cdot 797}{1852}} + \frac{\left(180 - \frac{495 \cdot 660}{1852}\right)^2}{\frac{495 \cdot 660}{1852}} + \frac{\left(252 - \frac{495 \cdot 395}{1852}\right)^2}{\frac{495 \cdot 395}{1852}}$$

$$= 585.9836$$

The p-value is:

$$p\text{-value} = P_{2 \cdot 2}(\chi^2 > 585.9836) = P_4(\chi^2 > 585.9836) \approx 0$$

i.e. the p-value is equal to or less than α; i.e. we reject the null hypothesis; i.e. the party identification of students and their parents depends on each other.

We reject the null hypothesis H_0 if and only if the observed value of the test statistic χ^2 is larger than the 95 percent point of the χ^2-distribution with $(I-1)(J-1)$ degrees of freedom, where I is the number of rows and J is the number of columns in the contingency table. The χ^2-distribution is not the exact distribution but an asymptotic distribution. The chi-square approximation is quite good provided that all values $\frac{n_{i\bullet} \cdot n_{\bullet j}}{n}$ are equal or larger than five.

In case of that the above rule of thumb is violated, it is often appropriate to combine two adjacent rows or columns. It is important to respect the rule of thumb! Cochran (1954) suggested another rule of thumb (cf. Agresti [1990] p. 246): *"Cochran studied the chi-square approximation for χ^2 in a series of articles. In 1954, he suggested that to test independence with df > 1, a minimum expected value of 1 is permissible as long as no more than about 20% of the cells have expected values below 5."*

Yates (1934) suggested a continuity correction in case of df = 1 (cf. Agresti page 68):

"For 2 × 2 tables, Yates (1934) suggested a correction to the Pearson statistic, ..., to adjust for using continuous chi-square distribution to approximate a discrete distribution. The corrected statistic gives *p*-values (from the chi-square distribution) that better approximate ..."

Result: We get the following rule of thumb:

Rule of thumb:

1) If df=1 the *p*-value is calculated with the continuity correction of Yates.

2) Up to 20 % of the cells could have expected counts $\dfrac{n_{i\bullet} \cdot n_{\bullet j}}{n}$ less than five.

3) The minimum of all expected counts is not smaller than one.

Remark: The χ^2-test of independence does not detect the direction of the relationship. Ordering of the categories is not taken into consideration.

Example 3.3 (*Party_Id.sav* c.f. Handl S. 9/10)
We want to investigate whether the subject of study affects the voting:

H_0 : Party identification and subject of study are stochastically independent
H_1 : Party identification and subject of study are dependent

One a basis of an observed sample we test the hypotheses. In an opinion survey we regard 20 female students and 80 male students:

female students		
Subject of study	party identification	
	CDU	SPD
BWL	4	12
VWL	2	2

male students		
Subject of study	party identification	
	CDU	SPD
BWL	46	24
VWL	4	6

In addition to the two variables X="Wahlverhalten (Voting)" and Y="Studienfach (Subject of study)", we want to examine for stochastic independence, there is a third variable Z="Geschlecht (Gender)". In order to take gender into account, a third variable can be entered as a **layer** in SPSS:

Layer = Gender

But we neglect the gender of the students so we get:

	party identification	
subject of study	CDU	SPD
BWL	50	36
VWL	6	8

Summing across the rows $n_{i\bullet}$ and summing across the columns $n_{\bullet j}$ we get:

	party identification		
subject of study	CDU	SPD	Σ
BWL	50	36	86
VWL	6	8	14
Σ	56	44	100

With the continuity correction of Yates we get the observed value of the test statistic as follows:

18

$$\chi^2_{\text{emp.}} = \sum_{i=1}^{I}\sum_{j=1}^{J}\frac{\left(\left|\,n_{ij}-\frac{n_i.\cdot n_{.j}}{n}\,\right|-0.5\right)^2}{\frac{n_i.\cdot n_{.j}}{n}}$$

$$= \frac{\left(\left|\,50-\frac{86\cdot56}{100}\,\right|-0.5\right)^2}{\frac{86\cdot56}{100}} + \frac{\left(\left|\,36-\frac{86\cdot44}{100}\,\right|-0.5\right)^2}{\frac{86\cdot44}{100}}$$

$$+\frac{\left(\left|\,6-\frac{14\cdot56}{100}\,\right|-0.5\right)^2}{\frac{14\cdot56}{100}} + \frac{\left(\left|\,8-\frac{14\cdot44}{100}\,\right|-0.5\right)^2}{\frac{14\cdot44}{100}}$$

$$= \frac{(|\,50-48.16\,|-0.5)^2}{48.16} + \frac{(|\,36-37.84\,|-0.5)^2}{37.84}$$

$$+\frac{(|\,6-7.84\,|-0.5)^2}{7.84} + \frac{\left(\left|\,8-\boxed{6.16}\,\right|-0.5\right)^2}{\boxed{6.16}}$$

$$= 0.03728405 + 0.04745243 + 0.2290306 + 0.2914935$$

$$= 0.6052606$$

$$\approx 0.6053$$

We are checking the rule of thumb:

1) The degree of freedom is one, df=1; though the p-value has to be computed with the continuity correction of Yates.

2) The minimum expected count is 6.16 (see the box in the computation of the empirical value of the test statistic) and larger than one.

3) All expected counts 48.16 and 37.84 and 7.84 and 6.16 are larger than five. Thus no cell has expected counts less than five.

Therefore the rule of thumb is fulfilled.
We get the p-value $=P_1(\chi^2 > 0.6053) = 0.437$.

Thus, we have:

$$p\text{-value}=0.437 > 0.05 = \alpha$$

i.e. we don't reject H_0; i.e. the variables "subject of study" and "party identification" are stochastically independent. This means that the party-identification of a BWL-student and of a VWL-student happens at random.

If the rule of thumb of the Pearson Chi-Square Test is not fulfilled, we should summarize some categories of the considered variable or we should class the values of the variable into different intervals. After that perhaps we can run the test.

Example 3.4 (*Kriminalitaet.sav*)
We want to check with the Pearson Chi-Square test whether murder and death penalty are stochastically independent:

H_0 : Murder and death penalty are stochastic independent
H_1 : Murder and death penalty are stochastic dependent

Error of the 1st kind: Mistaken rejection of H_0; i.e. the test does not detect that the relationship between murder and death penalty is at random.

Error of the second kind: Mistaken acceptance of H_0; i.e. the test does not detect that there is a relationship between murder and death penalty., i.e. that murder is depending on death penalty.

The rule of thumb is not fulfilled for the Pearson-Chi-Square test for the two variables X="Murder" and Y="Death Penalty": 90.9 % of all cells have expected counts less than five and the minimum expected count is 0.41. Though we class the rates of murder into three classes/intervals:

1. class = up to 3 murder per 100 000 inhabitants
2. class = more than 3 up to 6 murder per 100 000 inhabitants
3. class = more than 6 murder per 100 000 inhabitants

We call the new variable "Murder_class".

⚠ The level of the classed variable is no longer scale, but ordinal.

We run the Pearson Chi-Square test for the two variables "Death Penalty" and "Murder_class". The rule of thumb is fulfilled: No cell has expected counts less than five and the minimum expected count is 6.18. The p-value of 0.573 indicates that there is no relationship between death penalty and murder cases in a state.

Remark: If we class a scale leveled variable we always have a loss of information. In the year 2020 the German virologist Christian Drosten was critically criticized, because he has classed the age for an examination of a relationship between age and viral load of Covid-19.

3.2 Summary

The Pearson Chi-Square test verifies whether the two variables X and Y are stochastically independent based on a bivariate data set (x_1, y_1), (x_2, y_2), ..., (x_n, y_n). The test is applicable if the rule of thumb is fulfilled. If not you can try to summarize categories or values of the variables to get less more cells with frequencies closed to zero in the cross table.

For a multivariate data set $(x_1, y_1), z_1)$, (x_2, y_2, z_2), ..., (x_n, y_n, z_n) we may select one variable as a layer and check the independence for the two other variables.

3.3 SPSS Commands

3.3.1 Pearson Chi-Square Test

Input

We want to run a chi-square test for example 3.3. We type 1 as the subject BWL value and 2 as the subject VWL value. We type 1 as the party CDU value and 2 as the party SPD value:

No.	Subject of study	Party identification	
1	1	1	
⋮	⋮	⋮	50 Persons
50	1	1	
51	1	2	
⋮	⋮	⋮	36 Persons
86	1	2	
87	2	1	
⋮	⋮	⋮	6 Persons
92	2	1	
93	2	2	
⋮	⋮	⋮	8 Persons
100	2	2	

The rule of thumb $\dfrac{n_{i\bullet} \cdot n_{\bullet j}}{n} \geq 5$ is fulfilled.

Commands

1) Open the file "Party_Id.sav"

2) Analyze → Descriptive Statistics → Crosstabs

3) Row(s) = "Subject"
   Column(s) = "Party_Id"

4) Click "Statistics" and select "Chi-square". Click "Continue".

5) Click "OK".

Observing the continuity correction of Yates in case of df = 1 the p-value is listed in the second row in the column "Asymptotic Significance (2-tailed)" of the "Chi-square Test" table and is displayed as 0.437.

(In case of df > 1 the p-value is listed in the first row in the column "Asymptotic Significance (2-tailed)" of the "Chi-square Test" table.)

Output

Chi-Square Tests

	Value	df	Asympt. Sig. (2-tailed)	Exact Sig. (2-tailed)	Exact Sig. (1-tailed)
Pearson Chi-Square	1.141^b	1	.285		
Continuity Correction^a	.605	1	.437		
Likelihood Ratio	1.132	1	.287		
Fisher's Exact Test				.386	.218
Linear-by-Linear Association	1.130	1	.288		
N of Valid Cases	100				

a. Computed only for a 2x2-table

b. 0 cells (.0%) have expected count less than 5. The minimum expected count is 6.16.

3.3.2 How to Select Cases

We want to select only special cases of a variable in a data set. We discrimi-nate two cases: First, the variable is a string variable like X=Gender with the outcomes f=female, m=male and t=third. Secondly, the variable X=Gender is a numeric variable with the outcomes 1=female, 2=male, 3=third.
We want to select only women and men of the data set.
1st Case: Gender = string variable
⚠ Use of quotation marks

1) Data → Select Cases

2) Select "If condition is satisfied"
   Click "If ... "
   (Gender = "f") | (Gender= "m")
   Hint: The vertical line "|" is the symbol for the logical or.
   Continue

3) ok

Instead of choosing Gender="f" or Gender="m" you may also se-lect: Gender ~= "t" (not equal).

23

2nd Case: Gender = numeric variable

      1) Data → Select Cases

      2) Select "If condition is satisfied"
         Click "If ... "
         (Gender = 1) | (Gender = 2)
         Continue

      3) ok

Instead of choosing Gender="1" or Gender="2" you may also select: Gender ~= "3" (not equal).

3.3.3 How to Class Cases

SPSS offers two options to class a scale leveled variable. In the first option, we choose the lower and upper bound of a class ourselves. In the second option, we only specify how many classes we want to have and about the same number of cases then is lying in each class.

Example: In example 3.4 we want to have three classes of the variable murder of the file *Kriminalitaet.sav*.

1. Option:

1. Class = up to 3 murder per 100 000 inhabitants
2. Class = more than 3 up to 6 murder per 100 000 inhabitants
3. Class = more than 6 murder per 100 000 inhabitants

   1) Please open the file *Kriminalitaet.sav*.

   2) Transform → Recode into Different Variables ...

   3) Input Variable/Numeric Variable = Murder (*German: Mord*)

   4) Output Variable
      Name = Murder_class
      Change

   5) Old and New Values

6) Old Value
   Range, LOWEST through value: 3
   New Value
   Value = 1
   Add

7) Old Value
   Range: 3 through 6
   New Value
   Value = 2
   Add

8) Old value
   Range, value through HIGHEST: 6
   New Value
   Value = 3
   Add
   Continue

9) ok

The value 3 belongs to the class that was named first. The 1st class was named first as we have recoded the variable murder, though the value 3 belongs to the first class. If we had recoded the 2nd and 3rd class first, the value 3 would belong to the 2nd class.
Finally we should label the values of the lower and the upper bound of the intervals of the variable Murder_class as follows:

1) Select "Variable View".

2) In the column "Values " of the row variable Murder_class click on the light blue windowof the cell.

3) Value Labels
   Value = 1
   Label = up to 3
   Add

4) Value = 2
   Label = 4, 5 or 6
   Add

5) Value = 3
   Label = 7 or more
   Add

6) ok

2. Option:
We want to decompose the variable "Murder_class" in three classes. Every class should have about the same number of cases, i.e. about 33 % of all cases are belonging to each class respectively.

1) Please open the file *Kriminalitaet.sav.*

2) Transform → Rank Cases ...

3) Variable(s): = Murder

4) Click "Rank Types ...".
   Deselect the default case "Rank".
   Ntiles: 3
   Continue

5) ok

In Data View we get a new column with the ordinal leveled variable named "NMurder" (Percentile Group of Murder). The range of the variable "NMurder" are the values 1, 2 or 3, the numbers of the class of every case. The class boundaries of the intervals have to be searched for later. The upper bound of the 1st class is 3, the upper bound of the 2nd class is 6. About 39.2 % of all cases are belonging to the 1st class, 29.4 % to the 2nd class and 31.4 % to the 3rd class.

4 Measures of Association

Main purpose: We are looking for a measure of association of values of two variables in a bivariate data set.

4.1 Bravais-Pearson Correlation Coefficient

The two variables must be leveled as follows:

nominal:	no
ordinal:	no
scale:	yes

The **Bravais-Pearson correlation coefficient** r is a measure of association of values of two scale leveled variables in a bivariate data set. The measure r takes values between -1 and $+1$:

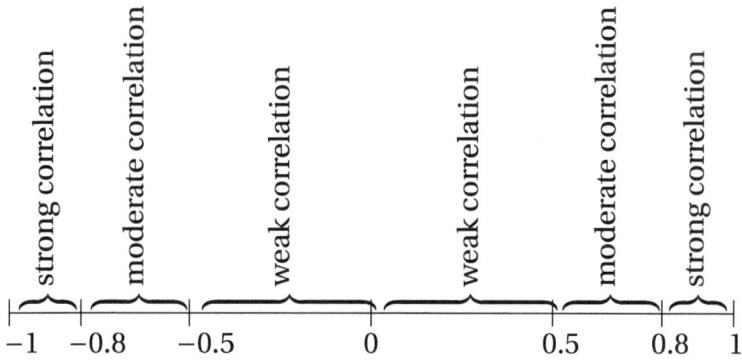

Example 4.1 (*Miles_Per_Gallon.sav* Berenson et al. [2015] p. 671)
We want to check whether there is a correlation between the gasoline mileage and the weight of a car. The correlation between the gasoline mileage

and the weight is -0.825, i.e. negative strong correlation. The gasoline mileage depends on the weight very well.

(You will find the data set in example 8.1.)

In the next example 4.2 we will learn more about the correlation coefficient of Bravais-Pearson.

Example 4.2 (Guessing Correlations)
In the internet please enter *http://istics.net/Correlations/*
Find the appropriate correlation values of each scatter plot.

If at least one of the two variables is not a scale leveled variable we are not allowed to use the correlation coefficient of Bravais-Pearson. So we have to look for other measures of association.

4.2 Spearman's rank correlation

The random variables must be leveled as follows:
        nominal: only dichotomous variables
        ordinal: yes
        scale: yes
A common measure of association for random variables X and Y is the rank correlation, or Spearman's correlation. The X values are ranked, and the observations are replaced by their ranks, similarly the Y observations are replaced by their ranks.

Example 4.3
For example the data set: $x_1, \ldots, x_4 = 36.7\ 28.1\ 53.4\ 47.6$
We sort the values in ascending order. The smallest value is designed with the rank 1, the second smallest value is designed with the rank 2 and so on. The largest value is designed with the rank 4. We get the following ranks of the data set:
        $R(x_1), \ldots, R(x_4) = 2\ 1\ 4\ 3$
If we have **ties** we have to use the so called **mid ranks**.
 Data set: $x_1, \ldots, x_5 = 36.7\ 28.1\ 53.4\ 47.6\ 28.1$
The smallest values are x_2 and x_5. They should have the ranks 1 and 2. The

arithmetic mean of 1 and 2 is 1.5. So x_2 and x_5 are designed with the mid rank 1.5. We get the following ranks of the data set:
$$R(x_1),\ldots,R(x_5) = 3 \quad 1.5 \quad 5 \quad 4 \quad 1.5$$
If we have several ties, we have to use the mid ranks, too.
Data set: $x_1,\ldots,x_6 = 36.7\ 28.1\ 36.7\ 28.1\ 53.4\ 28.1$
We get the following ranks of the data set:
$$R(x_1),\ldots,R(x_6) = 4.5 \quad 2 \quad 4.5 \quad 2 \quad 6 \quad 2$$

Let $R(X_i)$ denote the rank of X_i and $R(Y_i)$ the rank of Y_i. Using these paired ranks, the ordinary sample correlation is computed:

$$
\begin{aligned}
\rho &= \frac{\sum (R(x_i) - \overline{R}(x_i))(R(y_i - \overline{R}(y_i))}{\sqrt{\sum (R(x_i) - \overline{R}(x_i))^2} \cdot \sqrt{\sum (R(y_i) - \overline{R}(y_i))^2}} \\[2mm]
&= \frac{\sum R(x_i) \cdot R(y_i) - \frac{n}{4}(n+1)^2}{\sqrt{\sum R(x_i)^2 - \frac{n}{4}(n+1)^2} \cdot \sqrt{\sum R(y_i)^2 - \frac{n}{4}(n+1)^2}}
\end{aligned}
$$

This sample correlation Rho is called **Spearman's rank correlation** and is always lying in the interval $[-1;+1]$. Values of ρ near 1 are due to an association in the same direction. Values of ρ near -1 are due to an association in the opposite direction.
Let us consider an example.

Example 4.4 (*Umsatz_Zufried.sav*)
A salesperson has ten clients. The ten clients were asked about their sales X (in monetary units) and their satisfaction Y (1=dissatisfied, 2=satisfied, 3=very satisfied) with the salesperson. We get the following data set:

Sales	20.1	18.2	15.3	16.3	17.8	21.5	16.3	12.4	18.1	19.2
Satisf.	3	3	1	1	2	3	2	1	2	2

The salesperson wants to know whether there is a relationship between the amount of the sales and the degree of satisfaction. X="Sales" is a scale leveled variable and Y="Degree of Satisfaction" is an ordinal leveled variable. So it is not allowed to compute the correlation of Bravais-Pearson. But we are allowed to compute the rank correlation:

i	$R(x_i)$	$R(y_i)$	$R(x_i) \cdot R(y_i)$	$R(x_i)^2$	$R(y_i)^2$
	9	9	81	81	81
	7	9	63	49	81
	2	2	4	4	4
	3.5	2	7	12.25	4
	5	5.5	27.5	25	30.25
	10	9	90	100	81
	3.5	5.5	19.25	12.25	30.25
	1	2	2	1	4
	6	5.5	33	36	30.25
	8	5.5	44	64	30.25
Σ			370.75	384.5	376

Furthermore, we have: $\dfrac{n}{4}(n+1)^2 = \dfrac{10}{4} \cdot 11^2 = 302.5$

So we get:

$$\rho = \frac{370.75 - 302.5}{\sqrt{384.5 - 302.5} \cdot \sqrt{376 - 302.5}} = \frac{68.25}{\sqrt{82} \cdot \sqrt{73.5}} = 0.8791279$$

This means that there is a strong association in the same direction: large sales are going along with a high degree of satisfaction.

Remark: If we reverse the polarity of the variable Y=degree of satisfaction in example 4.4 as 1=very satisfied, 2=satisfied , 3=dissatisfied, the value of ρ would be negative, i.e. $\rho = -0.8791279$. The strong association would be in the opposite direction. Large sales are going along with a small degree of satisfaction. But small degrees of satisfaction are standing for satisfied or very satisfied. Thus the interpretation of $\rho = -0.880$ would be the same.

To comment ρ correctly it is useful to consider the cross table too. If $\rho > 0$, the cells top left are going along with the cells bottom right:

Sa-	Satisfaction		
les	very satisfied	...	dissatisfied
lower	☐		
⋮		☐	
high			☐

$$\rho = +0.879$$

If $\rho < 0$, the cells top right are going along with the cells bottom left:

Sa-	Satisfaction		
les	dissatisfied	...	very satisfied
lower			☐
⋮		☐	
high	☐		

$$\rho = -0.879$$

⚠ This is not a contradiction to the comment of r (increasing line $\Leftrightarrow r > 0$, decreasing line $\Leftrightarrow r < 0$), because for the comment of ρ (τ_b und γ) in a table, the coordinate system is folded down over the x-axis. Or in other words: If ρ resp. τ_b resp. γ are larger than zero, it applies both small values of X are going along with small values of Y as well large values of X are going along with large values of Y. If however ρ resp. τ_b resp. γ are smaller than zero, it applies both small values of X are going along with large values of Y as well large values of X are going along with small values of Y.

The rank correlation ρ between a dichotomous and an ordinal leveled variable is also used in psychology and designated as **biserial rank correlation**. For the special case that X is a dichotomous variable ($X = 1$, $X = 2$) and Y is an ordinal variable, the measure ρ is calculated as follows:

$$\rho = \frac{\frac{1}{12}\left(n^3 - n + 3n_1 n_2 n - C\right) - \sum_{i=1}^{n} d_i^2}{\sqrt{\frac{1}{12}n_1 n_2 n \left(n^3 - n - C\right)}}$$

where n is the sample size, n_1 is the total of the cases $X = 1$, n_2 is the total of cases $X = 2$, d_i is the difference of ranks rank(X_i) minus rank(Y_i) and $C = \sum_{i=1}^{b}(t_i^3 - t)$ mit $t_1, t_2, \ldots, t_b$ is the vector with the ties of Y, with t_i is the length of the ties respectively.

Example 4.5

Ten person were asked how important something is for them.

$\quad X =$ Gender 1=female, 2=male

$\quad Y =$ Degree of importance

$\qquad$ 1=very important

$\qquad$ 2=important

$\qquad$ 3=not important

$\qquad$ 4=completely unimportant

We get the following data set:

x_i	1	2	1	1	2	1	2	2	1	1
y_i	4	4	2	1	3	3	3	2	1	2

The rank correlation is: $\rho = 0.403$; i.e. for women, it is rather important, for men, is is rather unimportant.

We want to calculate ρ ourselves. The total of women is $n_1 = 6$, the total of men is $n_2 = 4$.

y_i	x_i	Rank(y_i)	Rank(x_i)	d_i	d_i^2
4	1	9.5	3.5	6	36
4	2	9.5	8.5	1	1
2	1	4	3.5	0.5	0.25
1	1	1.5	3.5	−2	4
3	2	7	8.5	−1.5	2.25
3	1	7	3.5	3.5	12.25
3	2	7	8.5	−1.5	2.25
2	2	4	8.5	−4.5	20.25
1	1	1.5	3.5	−2	4
2	1	4	3.5	0.5	0.25
Σ					82.25

The ties of Y are $(t_1, t_2, t_3, t_4) = (2, 3, 3, 2)$. (twice very important, three times important, three times not important, twice completely unimportant). Thus we get for C:

$$C = (2^3 - 2) + (3^3 - 3) + (3^3 - 3) + (2^3 - 2) = 6 + 24 + 24 + 6 = 60$$

The value of ρ is:

$$
\begin{aligned}
\rho &= \frac{\frac{1}{12}\left(10^3 - 10 + 3 \cdot 6 \cdot 4 \cdot 10 - 60\right) - 82.50}{\sqrt{\frac{1}{12} \cdot 6 \cdot 4 \cdot 10 \cdot (10^3 - 10 - 60)}} \\[2mm]
&= \frac{137.5 - 82.50}{\sqrt{18\,600}} \\[2mm]
&= \frac{55}{136.3818} \\[2mm]
&= 0.403
\end{aligned}
$$

So we get as well $\rho = 0.403$.

Remark: The rank correlation coefficient ρ gives based on a data set, how strong or weak the relationship between two variables X, Y is. However, ρ nothing says whether the x-values depend on Y or vice versa, the y-values depend on X. Which variable depends on which variable is unique and only the decision/interpretation of the statistician.

For the special case of two dichotomous variables X, Y the Spearman rank correlation coefficient equals the **Phi-Coefficient** Φ:
$$\rho(X, Y) = \Phi(X, Y).$$

4.3 Kendall's tau-b

The random variables must be leveled as follows:

nominal:	only dichotomous variables
ordinal:	yes
scale:	yes

First, we only want to measure the direction (positive or negative) but not the strength of the association between two variables.

Example 4.6 (*Einkommen_Job_Zufried.sav* Agresti [1990] p. 21)

X = Income (US$) (scale variable; note that income class is an ordinal variable)

Y = Job satisfaction (ordinal variable)

Does the level of the income affect the job satisfaction?

Asking 901 persons we get the following sample observations:

Income	Job Satisfaction			
	Very Dissatisfied	Little Dissatisfied	Moderately Satisfied	Very Satisfied
< 6000	20	24	80	82
6000 – 15000	22	38	104	125
15000 – 25000	13	28	81	113
> 25000	7	18	54	92

We assign four levels of income and four levels of job satisfaction:

Income		job satisfaction	
range	level	degree	level
< 6000	1	very dissatisfied	1
6000 – 15000	2	little dissatisfied	2
15000 – 25000	3	moderately satisfied	3
> 25000	4	very satisfied	4

For example, the sample observation $(x_a; y_a) = (2; 1)$ means that the corresponding person has an income between 6000 and 15000 US$ and is very dissatisfied with his job.

We want to analyze the direction (positive or negative) of the association between the level of income and the degree of job satisfaction. Perhaps, high income is going along with high job satisfaction. We use the designation concordance and discordance to check this association. A pair of bivariate sample observations (x_a, y_a) and (x_b, y_b) is called **concordant**, if and only if:

$$(x_a - x_b) \cdot (y_a - y_b) > 0$$

i.e. if person a has a higher income than person b and further a is more satisfied than b, then the pair (a, b) is called a concordant pair.

A pair of bivariate sample observations (x_a, y_a) and (x_b, y_b) is called **discordant**, if and only if:

$$(x_a - x_b) \cdot (y_a - y_b) < 0$$

i.e. if person a has a higher income than person b, but a is less satisfied than b, then the pair (a, b) is called a discordant pair.

The pair is **tied** if x_a, x_b and/or y_a, y_b have the same classification:

$$(x_a - x_b) \cdot (y_a - y_b) = 0.$$

The total number C of concordant pairs, denoted by C, equals:

$$
\begin{aligned}
C \quad = \quad & 20(38 + 104 + 125 + 28 + 81 + 113 + 18 + 54 + 92) \\
& +24(104 + 125 + 81 + 113 + 54 + 92) \\
& +80(125 + 113 + 92) \\
& +22(28 + 81 + 113 + 18 + 54 + 92) \\
& +38(81 + 113 + 54 + 92) \\
& +104(113 + 92) \\
& +13(18 + 54 + 92) \\
& +28(54 + 92) \\
& +81 \cdot 92 \\
= \quad & 109\,520
\end{aligned}
$$

The number D of discordant pairs of observations is:

$$\begin{aligned}
D \quad = \quad & 24(22+13+7) \\
& +80(22+38+13+28+7+18) \\
& +82(22+38+104+13+28+81+7+18+54) \\
& +38(13+7) \\
& +104(13+28+7+18) \\
& +125(13+28+81+7+18+54) \\
& +28 \cdot 7 \\
& +81(7+18) \\
& +113(7+18+54) \\
= \quad & 84\,915
\end{aligned}$$

In this example, the number of concordant pairs is exceeding the number of discordant pairs.

The difference $\tau = C - D$ (*read: tau*) is denoted as **Kendall's tau**.

We get $\tau = 109\,520 - 84\,915 = 24\,605$, that suggests a tendency for low income to occur with low job satisfaction and high income with high job satisfaction.
A positive value of τ will be interpreted as a positive relationship between the two variables. A negative value of τ will be interpreted as a negative relationship between the two variables. However, the actual level of τ does not have much meaning. We know nothing about the strength of the relationship.

In the observation sample the frequencies of income are:

X	< 6000	$6000 - 15000$	$15000 - 25000$	> 25000
n_i^X	206	289	235	171

In the observation sample the frequencies of the degrees of job satisfaction are:

Y	very dissat.	lit. dissat.	mod. sat.	very sat.
n_j^Y	62	108	319	412

We want to standardize the measure τ. Therefore we need the following values:

$$T_X = \sum_i n_i^X (n_i^X - 1)/2$$
$$= \frac{1}{2}(206 \cdot 205 + 289 \cdot 288 + 235 \cdot 234 + 171 \cdot 170)$$
$$= 104\,761$$

$$T_Y = \sum_j n_j^Y (n_j^Y - 1)/2 = \frac{1}{2} \cdot 286\,112 = 143\,056$$

We standardize the measure τ:

$$\tau_b = \frac{C - D}{\sqrt{(\frac{n}{2}(n-1) - T_X)(\frac{n}{2}(n-1) - T_Y)}}$$

This measure is denoted as **Kendall's tau-b**. The measure τ_b ranges between -1 and $+1$. A value of τ_b near ± 1 means that the association is strong. A value of τ_b near 0 means that the association is rather weak. The value $\tau_b = -1$ indicates perfect disagreement, the value $\tau_b = 0$ indicates no agreement, the value $\tau_b = 1$ indicates perfect agreement.

We get:

$$\tau_b = \frac{109\,520 - 84\,915}{\sqrt{(405\,450 - 104\,761)(405\,450 - 143\,056)}} = 0.08759668054$$

i.e. $\tau_b \approx 0.088$; i.e. there is a weak association between the variables.
The rank correlation coefficient ρ of Spearman is a suitable measure for relationship too. The value is lager: $\rho = 0.102$; i.e. the correlation is weak. I.e. the level of income has little impact on the degree of job satisfaction.

Remark: A reversal in the category orderings in example 4.6 of the variable degree of satisfaction:

Income	Satisfaction			
	very satisfied	satis- fied	dis- satisfied	very dis- satisfied
< 6000	82	80	24	20
6 000 − 15 000	125	104	38	22
15 000 − 25 000	113	81	28	13
> 25 000	92	54	18	7

simply causes a change in the sign of τ_b. We get $\tau_b = -0.088$. To get a reversal in the category orderings of the variable satisfaction we have to recode the data.

Only for quadratic crosstabs the values of Kendall-tau-b are in the closed interval $[-1; +1]$, for all non-quadratic crosstabs the values of Kendall-tau-b are in the open interval $(-1; +1)$, invariably. Around to compensate for this difference became another standardization made by Kendall-tau and this measure is denoted by **Kendall-Tau-c**.

$$\tau_c = (C - D) \cdot \frac{2m}{n^2(m-1)} = 24\,605 \cdot \frac{2 \cdot 4}{901^2 \cdot 3} = 0.080\,824\,4$$

where m is the number of rows or columns whichever is less. And n denotes the sample size. For all crosstabs, the values of Kendall-Tau-b are in the closed interval $[-1; +1]$. The disadvantage of the measure Kendall-Tau-c is, that τ_c depends on the dimension of the crosstab.

In the example 4.7 we can see this disadvantage.

Example 4.7
For the variable X = "Number of kids in a family" and for the variable Y = "Density of population at the residence of the family" with the levels $Y = 1, 2, 3, 4$ we have ask $n = 12$ families:

X	Y			
	1	2	3	4
1	2	2	1	1
3	1	1	2	2

We get a 2 × 4-cross table and the following measures:

$$\tau_b = \frac{12}{\sqrt{36 \cdot 54}} = 0.272\,165\,5$$

$$\tau_c = 12 \cdot \frac{2 \cdot 2}{12^2 \cdot 1} = \frac{1}{3} = 0.\overline{3}$$

If we change the outfit of the cross table as follows:

X	Y 1	2	3	4
1	2	2	1	1
2	0	0	0	0
3	1	1	2	2

We get a 3 × 4-cross table and the measures:

$$\tau_b = \frac{12}{\sqrt{36 \cdot 54}} = 0.272\,165\,5$$

$$\tau_c = 12 \cdot \frac{2 \cdot 3}{12^2 \cdot 2} = \frac{1}{4} = 0.25$$

This means that for both cross tables the values of τ_b are the same, but the values of τ_c differs.

Kendall-Tau-c is disputed among statisticians.

4.4   Gamma Coefficient

The two random variables must be leveled as follows:

        nominal: only dichotomous variables

        ordinal: yes

        scale: yes

More sensitive (c.f. Agresti[2002] p. 68) than Kendall's Tau-b as a further measure of association is the so called **gamma coefficient**:

$$\gamma = \frac{C-D}{C+D} \in [-1; +1]$$

We consider once more again the example 4.6.

Example 4.8 (*Einkommen_Job_Zufried.sav* c.f. Agresti [1990] p. 21)
X = Income-class (US\$) (ordinal variable); category 1 = less than 6 000 US\$, category 2= between 6 000 but and 15 000 US\$, between 15 000 and 25 000; category 4= over 25 000 US\$
Y = Job satisfaction (ordinal variable); 1= very dissatisfied, 2=dissatisfied, 3=satisfied, 4=very satisfied
Does the level of the income affect the job satisfaction?
Asking 901 persons we get the following sample observations:

	Job Satisfaction			
Income	Very Dissatisfied	Little Dissatisfied	Moderately Satisfied	Very Satisfied
< 6 000	20	24	80	82
6 000 – 15 000	22	38	104	125
15 000 – 25 000	13	28	81	113
> 25 000	7	18	54	92

The number of concordant pairs, denoted by C, equals $C = 109\,520$. The number of discordant pairs of observations is $D = 84\,915$. We obtain:

$$\gamma = \frac{109\,520 - 84\,915}{109\,520 + 84\,915} = 0.126546 \approx 0.127$$

This means that there is a weak association of monotonicity: more income is going along with a higher degree of satisfaction.

A reversal in the category orderings of one variable simply causes a change in the sign of γ.

4.5 Coefficient of Contingency

Up to know we have no measure of association in case of at least one nominal leveled variable with three or more categories.

We get such a measure with the coefficient of contingency C, but with the disadvantage that C cannot measure any direction of the association.

The two random variables must be leveled as follows:

nominal: yes
ordinal: yes
scale: yes

The strength of association between the two variables can be assessed by the **coefficient C of contingency**:

$$C = \left(\frac{\chi^2}{\chi^2 + n} \right)^{0.5} \in [0; 1)$$

where χ^2 is the empirical value of the Chi-Square teststatistic:

$$\chi^2 = \sum_{i=1}^{I} \sum_{j=1}^{J} \frac{\left(n_{ij} - \frac{n_{i\cdot} \cdot n_{\cdot j}}{n} \right)^2}{\frac{n_{i\cdot} \cdot n_{\cdot j}}{n}}$$

Values of C near zero indicate a weak or no relationship, values of C near +1 indicate a strong relationship between the two variables:

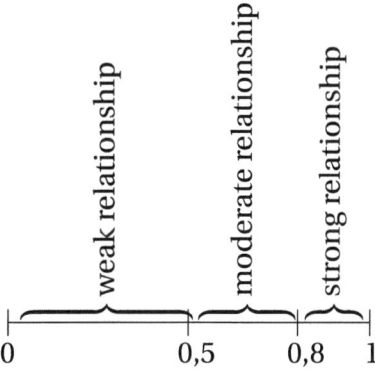

Example 4.9 (*customer_base.sav* c.f. Tutorial of SPSS)
Is there an association between the job category (Managerial and Profes-
sional, Sales and Office, Service, Agriculture and Natural Resources, Preci-
sion Production, Fabrication) and the job satisfaction?

Job cate- gory	Job Satisfaction					Total
	Highly dissatisfied	Somewhat dissatisfied	Neutral	Somewhat satisfied	Highly satisfied	
Managerial and Professional	245	332	319	274	209	1379
Sales and Office	472	378	347	262	171	1630
Service	85	114	143	142	144	628
Agricultural and Natural Resources	30	38	48	52	50	218
Precision Produc-, tion, Craft, Repair	55	65	86	119	121	446
Operation, Fabrica-, tion, General Labor	80	114	149	165	191	699
Total	967	1041	1092	1014	886	5000

The empirical value of the chi-square test statistic is:

$$\chi^2 = \frac{\left(245 - \frac{1379 \cdot 967}{5000}\right)^2}{\frac{1379 \cdot 967}{5000}} + \ldots + \frac{\left(191 - \frac{699 \cdot 886}{5000}\right)^2}{\frac{699 \cdot 886}{5000}} = 315.982$$

The coefficient of contingency is:

$$C = \left(\frac{315.982}{315.982 + 5000}\right)^{0.5} = 0.243803 \approx 0.244$$

This means that there is only a weak association between job category and
the degree of job satisfaction.

In addition to the coefficient of contingency we get with the empirical value
χ^2 of the test statistic of Pearson's test of independence another measure of
association:

Cramérs $V = \left(\dfrac{\chi^2}{n(k-1)}\right)^{0,5} \in [0;1]$, with $k = \min \{I, J\}$.

4.6 Summary

Measures of the strength (weak, moderate, strong) of the association of the x-values and y-values of a bivariate data set $(x_1, y_1), (x_2, y_2), \ldots, (x_n, y_n)$ are:

Measure	Level of X and Y	Direction
r of Pearson	scale	yes
τ_b of Kendall	dichotomous, ordinal, scale	yes
ρ von Spearman	dichotomous, ordinal, scale	yes
γ Gamma	dichotomous, ordinal, scale	yes
C Contingency coefficient	nominal, ordinal, scale	no

Except the coefficient of contingency C the sign of all other measures indicates the direction (positive or negative) of the association.

In case of a multivariate data set we can use in SPSS a third variable as a layer.

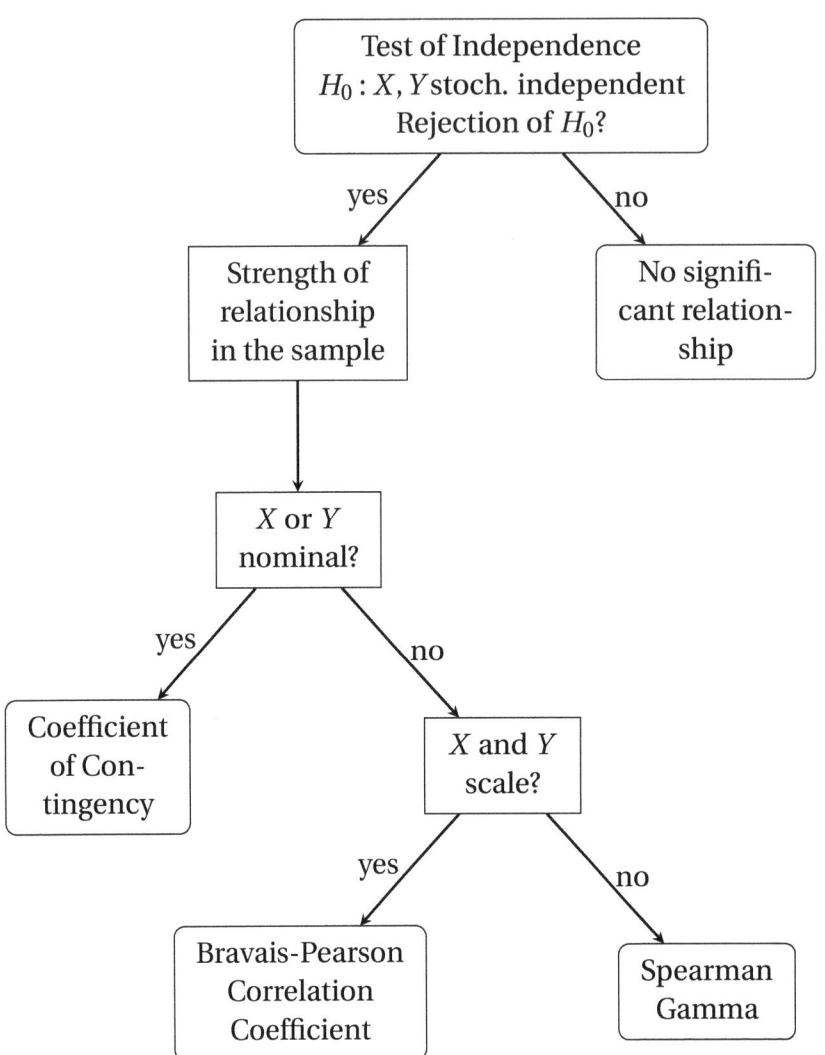

4.7 SPSS Commands

4.7.1 Recoding data

We want to recode the degree of satisfaction in example 4.4 in reverse order:

1) Transform → Recode into different Variables . . .

2) Input-Variable=Job_Satisf
   Output -Variable =Job_Satisf_opposite
   Change

3) Old and New Values
   Old Value = 4   New Value = 1   Add
   Old Value = 3   New Value = 2   Add
   Old Value = 2   New Value = 3   Add
   Old Value = 1   New Value = 4   Add
   Continue

4) Ok

The values of the new variable Satisf_opposite are shown in the last column in the Data View.

4.7.2 Correlation Coefficient of Bravais-Pearson

We compute the correlation of Bravais-Pearson between mileage and weight of example 4.1:

1) Analyze → Correlate → Bivariate . . .

2) Variables = MPG and weight
   Correlation Coefficients = Pearson

3) Click "ok".

The correlation coefficient between gasoline mileage and weight is -0.825.

4.7.3 Spearman's-rho

We compute the rank correlation between sales and degree of satisfaction of example 4.4:

Input

	Sales	Satisfaction
1	20,10	3,00
2	18,20	3,00
3	15,30	1,00
4	16,30	1,00
5	17,80	2,00
6	21,50	3,00
7	16,30	2,00
8	12,40	1,00
9	18,10	2,00
10	19,20	2,00

Commands

1. Option:

1) Open the file Umsatz_Zufried.sav

2) Analyze → Descriptive Statistics → Crosstabs ...

3) Row(s) ="Umsatz"
   Column(s) ="Zufried"

4) Click "Statistics". Select "Correlations".
   Click "Continue".

5) Click "OK".

These selections generate a cross table to comment the sign of Spearman-Rho, and a table with symmetric measures. The value of Spearman-Rho is 0.879.

Output

Symmetric Measures

		Value	Asymptotic Standard-Error[a]	Approximate t^b	Approximate Significance
Interval by Interval	Pearson's R	,831	,063	4,218	,003[c]
Ordinal by Ordinal	Spearman Correlation	,879	,066	5,217	,001[c]
N of Valid Cases		10			

a Not assuming the null hypothesis.
b Using the asymptotic standard error assuming the null hypothesis.
c Based on normal approximation.

2. Option:

1) Open the file "Umsatz_Zufried.sav "

2) Analyze → Correlate → Bivariate

3) Variables = "Umsatz"
         "Zufriedenheit"
   In "Correlation Coefficients" select "Spearman".

4) Click "OK".
   You will find the value 0.879 of the rank correlation in the SPSS-Output-Table "Correlations" in the row "Spearman-Rho".

Output

Correlations

			Umsatz	Zufried
Spearman-Rho	Umsatz	Correlation Coefficient	1.000	.879**
		Sig. (2-tailed)	.	.001
		N	10	10
	Zufried	Correlation Coefficient	.879**	1.000
		Sig. (2-taled)	.001	.
		N	10	10

** Correlation is significant at the 0.01 level (2-tailed).

4.7.4   Kendall's tau-b

We want to compute Kendall's tau-b between income and degree of job satisfaction of example 4.6:

Input

We have assigned four levels 1,2,3,4 of the variable income and of the variable job satisfaction:

No	Income	Job-Satisf.	
1	1	1	
⋮	⋮	⋮	20 persons
20	1	1	
21	1	2	
⋮	⋮	⋮	24 persons
44	1	2	
45	1	3	
⋮	⋮	⋮	80 persons
124	1	3	
125	1	4	
⋮	⋮	⋮	82 persons
206	1	4	
⋮	⋮	⋮	
⋮	⋮	⋮	
⋮	⋮	⋮	
731	4	1	
⋮	⋮	⋮	7 persons
737	4	1	
738	4	2	
⋮	⋮	⋮	18 persons
755	4	2	
756	4	3	
⋮	⋮	⋮	54 persons
809	4	3	
810	4	4	
⋮	⋮	⋮	92 persons
901	4	4	

Commands

1.Option:

1) Open the file Einkommen_Job_Zufried.sav

2) Analyze → Descriptive Statistics → Crosstabs ...

3) Row(s) ="Einkommen"
   Column(s) ="Job_Zufried"

4) Click "Statistics". Select "Kendall's tau-b".
   Click "Continue".

5) Click "OK".
   These selections produce a crosstabulation table and a table with symmetric measures. The value of Kendall's tau-b is 0.088.

Output

Income * Job_Satisfaction Crosstabulation

Count

		Job_Satisfaction				Total
		very dis-satis-fied	dis-satis-fied	satis-fied	very satis-fied	
In-come	< 6 000	20	24	80	82	206
	6 000 - 15 000	22	38	104	125	289
	15 000 - 25 000	13	28	81	113	235
	> 25 000	7	18	54	92	171
Total		62	108	319	412	901

This contingency table/cross table is very important for the interpretation of the sign of Kendall-Tau-b. In case of a positive sign (as here), there is a connection between top left in the contingency table to bottom right in the contingency table; i.e. low income is going along with little satisfaction and high income is going along with great satisfaction. However, the relationship is only weak with $\tau_b = 0.088$:

Symmetric Measures

		Value	Asymp. Std. Error^a	Approx. T^b	Approx. Sig.
Ordinal by Ordinal	Kendall's tau-b	.088	.028	3.091	.002
N of Valid Cases		901			

a. Not assuming the null hypothesis.
b. Using the asymptotic standard error assuming the null hypothesis.

2. Option:

1) Open the file Einkommen_Job_Zufried.sav "

2) Analyze → Correlate → Bivariate

3) Variables = "Einkommen"
   "Zufried"
   In "Correlation Coefficients" select "Kendall's tau-b".

4) Click "OK".
   You will find the value 0.088 of the rank correlation in the SPSS-Qutput-Table "Correlations" in the row "Kendall's tau_b".

Output

Correlations

			Income	Job_Satisf
Kendall's tau-b	Income	Correlation Coefficient	1.000	.088**
		Sig. (2-tailed)	.	.002
		N	901	901
	Job_Satisf	Correlation Coefficient	.088**	1.000
		Sig. (2-taled)	.002	.
		N	901	901

** Correlation is significant at the 0.01 level (2-tailed).

4.7.5 Gamma

We want to compute gamma between income and degree of job satisfaction of example 4.6:

Commands

1) Open the file Einkommen_Job_Zufried.sav

2) Analyze → Descriptive Statistics → Crosstabs ...

3) Row(s) ="Einkommen"
   Column(s) ="Zufried"

4) Click "Statistics". Select "Gamma".
   Click "Continue".

5) Click "OK".
   These selections produce a cross tabulation table and a table with symmetric measures. The value of γ is 0.127.

Output

Symmetric Measures

		Value	Asymp. Std. Error[a]	Approx. T[b]	Approx. Sig.
Ordinal by Ordinal	Gamma	.127	.041	3.091	.002
N of Valid Cases		901			

a. Not assuming the null hypothesis.

b. Using the asymptotic standard error assuming the null hypothesis.

4.7.6 Coefficient of Contingency

We compute the coefficient of contingency between job category and job satisfaction of the example 4.9:

Commands

1) Open the file "*customer_dbase.sav*"

2) Analyze → Descriptive Statistics → Crosstabs

51

3) Row(s) ="Job category"
   Column(s) ="Job satisfaction"

4) Click "Statistics" and select "Contingency coefficient". Click "Continue".

5) Click "OK"

The value of the coefficient of contingency is 0.244 (weak relationship).

Output

Symmetric Measures

		Value	Approx. Sig.
Nominal by Nominal	Contingency Coefficient	0.244	0.000
N of Valid Cases		5000	

5 Goodness-of-Fit Test Normal Distribution

Main purpose: We want to state that a distribution of a population belongs to the Normal distributions. The random variable must be leveled as follows:

nominal:	no
ordinal:	no
scale:	yes

In many situations it is assumed that the population belongs to the Normal distributions. For "large" sample sizes ($n \geq 30$) the Central-limit Theorem guarantees the assumption of Normal distribution in many cases. For small sample sizes $n < 30$ we will check the Normal distribution with a statistical test.

If the assumption of a Normal distribution is fulfilled, we are able to calculate probabilities of special events. For example we can calculate the probablitiy of a loss at the stock market.

Example 5.1
The probability of loss of 2 000 € after one year if we invest 20 000 € at the Germanstock market was 70% before the financial crisis in the year 2008. This probability can be calculated based on the last years of the Dax values.

In SPSS there are several ways to check Normal distributions:

- Lilliefors test

- Shapiro-Wilk test

- with a glance at the histogram

- with a glance at the empirical cumulative distribution function

- with a glance at the Quantile-Quantile Plot

There is the Jarque-Bera test, too. For the test decision the Jarque-Bera test calculates the sample skewness and the sample curtosis. SPSS does not offer the Jarque-Bera test.

A test is a better statistical inference than a (subjective) visual check of a diagram.

First, we are interested in testing.

5.1 Lilliefors Test

The Russian mathematicians Kolmogorov (1933) and Smirnov suggested a test (KS-test) to detect whether a population belongs to Normal distributions $N(\mu; \sigma)$ with known parameters μ and σ. Lilliefors (1967) improved the KS-test in case of unknown parameters. He estimates the unknown parameters with a sample mean $\bar{x}$ and a sample variance s^2.

Example 5.2 (*Spar_Guthaben.sav*)
Are savings balances (in German: Sparguthaben) Normally distributed?

We are interested in testing at level $\alpha = 0.05$ whether the random variable X = "savings balance (in €) of a client" is Normally distributed: H_0 : "The distribution of X is the Normal distribution; brief: $F_X = N$" versus H_1 : "The distribution of X is not the Normal distribution; brief: $F_X \neq N$.

We reject the null hypothesis if and only if there is a "big" difference between the empirical cumulative distribution function and the Normal cumulative distribution function; this means that if the difference is larger than the value of c_n:

n	c_n	n	c_n
4	.381	14	.227
5	.337	15	.220
6	.319	16	.213
7	.300	17	.206
8	.285	18	.200
9	.271	19	.195
10	.258	19	.190
11	.249	25	.180
12	.242	30	.161
13	.234		

$$n > 30 \qquad c_n = \frac{.886}{\sqrt{n}}$$

We are testing on the basis of observed sample values. The following savings balances of $n = 12$ clients are observed:

9 800 9 300 15 200 8 600 12 200 11 600
10 200 8 700 6 900 9 600 15 500 7 200

The estimators are the average value $\bar{x}$ and the empirical standard deviation s:

$\bar{x} = 10400$ is the estimator for μ
$s = \sqrt{s^2} = 2773.2488$ is the estimator for σ

Let $x_{(i)}$ denote the ordered sample values. And let $z_{(i)} = \frac{x_{(i)} - 10400}{2773.25}$ denote the standardized sample values. The values of the empirical cumulative distribution function are $F(x_{(i)}) = \frac{i}{12}$. And the corresponding values of the Normal distribution function are $F(X \leq x_{(i)}) = F_U(z_{(i)})$. We calculate the maximum difference $\frac{i}{n} - F_U(z_{(i)})$ between the empirical cumulative distribution function and the Normal cumulative distribution function. After this, we calculate the maximum difference $F_U(z_{(i)}) - \frac{i-1}{12}$ between the Normal cumulative distribution function and the empirical cumulative distribution function:

55

$x_{(i)}$	$z_{(i)}$	$F(x_{(i)}) = \frac{i}{12}$	$F_U(z_{(i)})$	$\frac{i}{12} - F_U(z_{(i)})$	$F_U(z_{(i)}) - \frac{i-1}{12}$
6900	−1.26	.0833	.1038	−.0205	.1038
7200	−1.15	.1667	.1251	.0416	.0418
8600	−0.65	.2500	.2578	.0078	.0911
8700	−0.61	.3333	.2709	.0624	.0209
9300	−0.40	.4167	.3446	.0721	.0113
9600	−0.29	.5000	.3858	.1141	−.0308
9800	−0.22	.5833	.4129	.1704	−.0871
10200	−0.07	.6667	.4721	.1946	−.1112
11600	0.43	.7500	.6664	.0836	−.0003
12200	0.65	.8333	.7422	.0911	−.0078
15200	1.73	.9167	.9582	.0415	.1249
15500	1.84	1.0000	.9671	.0329	.0504

The maximum difference between the empirical cumulative distribution function and the Normal distribution function is 0.1946 (5th column). The maximum difference between the Normal cumulative distribution function and the empirical cumulative distribution function is 0.1249 (last column). Thus, the maximum difference is 0.1946 and that is the observed value of the test statistic.

For $n = 12$ the level $\alpha = 0.05$ test rejects the null hypothesis if and only if the maximum difference is larger than the critical value 0.242. But the value 0.1946 of the maximum difference is less than the critical value thus we don't reject the null hypothesis.
i.e. the data do not indicate that the hypothesis that the savings balance of a client are Normally distributed should be rejected.

Using SPSS we compare the calculated p-value with $\alpha = 0.05$ to get the test decision:

Example 5.3 (*Spar_Guthaben.sav*)

The lower bound of the p-value in example 5.2 is 0.2; this means p-value $\geq$ 0.2; i.e. no rejection of H_0; thus savings balances of a client have Normal distribution.

The Lilliefors test performs well for small sample sizes $n \leq 50$. For larger sample sizes often the Lilliefors test flops, even if the sample was generated from a Normal distribution.

5.2   Shapiro-Wilk Test

In the year 1965 Shapiro and Wilk developed another goodness-of-fit test for the Normal distribution. The test is labeled the Shapiro-Wilk test. Some studies have concluded that this test has larger power than the Lilliefors test in many situations:

Example 5.4 (*Spar_Guthaben.sav*)

With a level $\alpha = 0{,}05$ test we want to check the Normal distribution of savings balances based on a sample of size $n = 12$. The twelve observations are listed in example 5.2.

SPSS calculates the p-value as 0.199; so we should not reject H_0. Thus savings balances are Normally distributed.

The Shapiro-Wilk test performs well even for small sample sizes ($n < 20$).

5.3   Jarque-Bera Test

In the year 1980 Anil K. Bera and Carlos M. Jarque proposed another goodness-of-fit test to detect a Normal distribution. For the test decision the empirical skewness S (a measure of the symmetry or asymmetry of a distribution):

$$S = \frac{\frac{1}{n}\sum_{i=1}^{n}(x_i - \bar{x})^3}{\left(\frac{1}{n}\sum_{i=1}^{n}(x_i - \bar{x})^2\right)^{1.5}}$$

and the empirical **kurtosis** K (a measure of the tailweight of a distribution):

$$K = \frac{\frac{1}{n}\sum_{i=1}^{n}(x_i - \bar{x})^4}{\left(\frac{1}{n}\sum_{i=1}^{n}(x_i - \bar{x})^2\right)^2}$$

are needed.

Jarque-Bera test
H_0:   The distribution is the Normal distribution; brief: $F_X = \mathsf{N}$
versus
H_1:   The distribution is not the Normal distribution; brief: $F_X \neq \mathsf{N}$
Rejection of $H_0 \Leftrightarrow p$-value ≤ 0.05

The Jarque-Bera test is not listed in the SPSS-procedures. So we have to compute the so called critical values by ourselves. The Jarque-Bera test statistic T of the sample is:

$$T = \frac{n}{6}\left(S^2 + \frac{(K-3)^2}{4}\right)$$

The measure $K - 3$ is labeled the **excess**. Symmetric distributions like the Normal distribution can be shown to have $S = 0$:

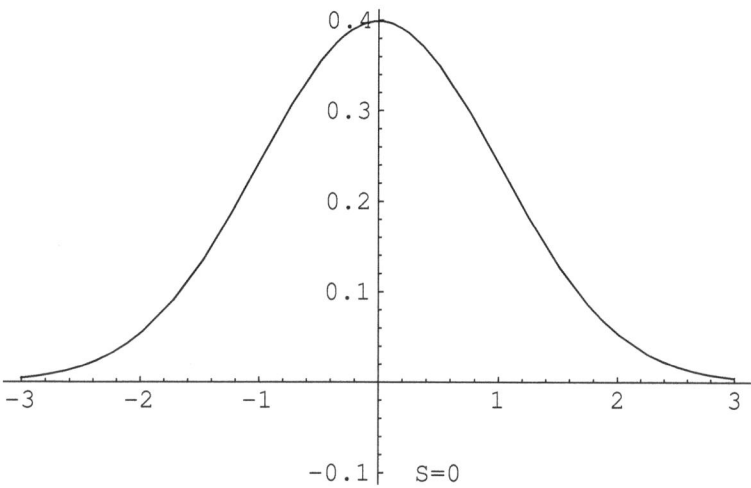

Asymmetric distributions with a shape skewed to the right (left steep) can be shown to have positive values of S:

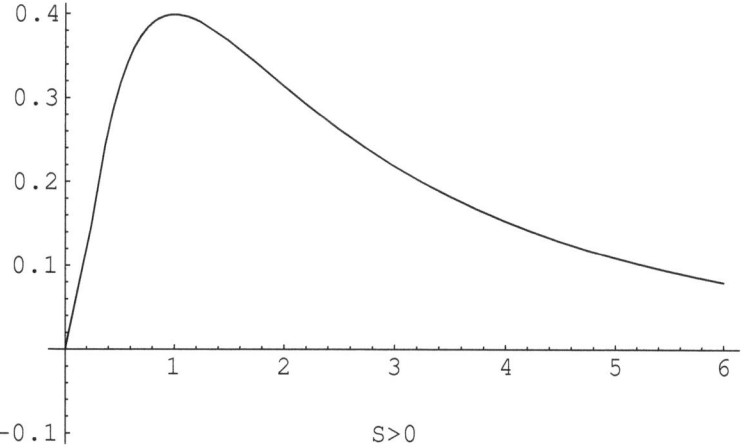

Asymmetric distributions with a shape skewed to the left (right steep) can be shown to have negative values of S:

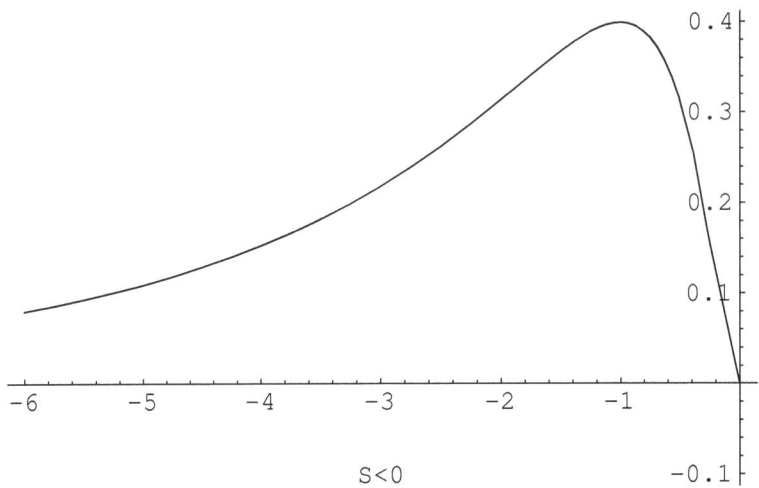

S<0

The kurtosis is used as a measure of the degree of flatness of a density near its center. The Normal distribution can be shown to have $K = 3$. Values larger than three indicate that a density is more peaked around its center than the density of the standard Normal curve. Values less than three indicate that a density is more flat around its center than the density of the standard Normal curve.

The test statistic T has an asymptotic chi-square distribution with two degrees of freedom. The test decision is as follows:

α	Rejection of H_0
0.01	$T > 9.21034$
0.05	$T > 5.991465$
0.10	$T > 4.60517$

Because the Jarque-Bera test is not listed by SPSS we have to compute the test statistic by ourselves.

Example 5.5 (*Spar_Guthaben.sav*)
We like to check whether saving balances of a client (in Euro) are Normally distributed. The data set of size $n = 12$ is:

9 800 9 300 15 200 8 600 12 200 11 600

$$10\,200 \quad 8\,700 \quad 6\,900 \quad 9\,600 \quad 15\,500 \quad 7\,200$$

The first four moments are:

$$\bar{x} = 10\,400$$

$$\frac{1}{n} \sum_{i=1}^{n} (x_i - 10\,400)^2 = 7\,050\,000$$

$$\frac{1}{n} \sum_{i=1}^{n} (x_i - 10\,400)^3 = 13\,529\,000\,000$$

$$\frac{1}{n} \sum_{i=1}^{n} (x_i - 10\,400)^4 = 124\,642\,300\,000\,000$$

So we get for the empirical skewness:

$$S = \frac{13\,529\,000\,000}{7\,050\,000^{1.5}} = 0.7227399$$

The empirical kurtosis is as follows:

$$K = \frac{124\,642\,300\,000\,000}{7\,050\,000^2} = 2.507767$$

The test statistic T has the value:

$$T = \frac{12}{6} \left(0.7227399^2 + \frac{(2.507767 - 3)^2}{4} \right) = 1.165852$$

Hence we get $1.165852 < 5.991465$, so H_0 should not be rejected; this means that the data set belongs to a Normal distribution.

The software SPSS calculates the values of the skewness and the kurtosis as the following terms:

$$\text{SPSS-Skewness} \quad = \quad \frac{\sqrt{n(n-1)}}{n-2} \cdot S = \frac{\sqrt{12 \cdot 11}}{10} \cdot 0.7227399 = 0.830$$

$$\text{SPSS-Kurtosis} \quad = \quad \frac{(n-1)(n+1)}{(n-2)(n-3)} \cdot K - 3 \cdot \frac{(n-1)^2}{(n-2)(n-3)}$$

$$= \quad \frac{11 \cdot 13}{10 \cdot 9} \cdot 2.508 - 3 \cdot \frac{121}{10 \cdot 9} = -0.049$$

The so called critical values 9.21034, 5.991465 and 4.60517 of the Jarque-Bera test are inexact for small sample sizes ($n < 200$), because the approximation by the chi-square distribution does not work well. Monte Carlo simulations with 600 000 replications has evaluated the following critical values for $n = 20, 50, 100, 200$ (cf. Dong W. Cho and Kyung So Im [2002] and Büning [2007]):

n	critical values for		
	$\alpha = 0.01$	$\alpha = 0.05$	$\alpha = 0.10$
20	9.762	3.821	2.359
50	12.578	5.007	3.203
100	12.626	5.442	3.673
200	11.858	5.694	4.037
500	–	5.825	–
∞	9.210	5.991	4.605

Result: Do not use the Jarque-Bera test at level $\alpha = 0.05$ for sample sizes smaller than 200.

Remark: For the sample size n we get the values S and K for the Jarque Bera test from SPSS as follows:

$$S \quad = \quad \frac{n-2}{\sqrt{n(n-1)}} \cdot S_{SPSS} \approx S_{\text{SPSS}}$$

$$K \quad = \quad \frac{(n-2)(n-3)}{n^2 - 1} \cdot K_{SPSS} + 3 \cdot \frac{n-1}{n+1} \approx K_{\text{SPSS}} + 3$$

5.4   Histogram versus Normal Distribution

We plot two graphs simultaneously, the histogram and the Normal curve, to check visually whether the observed sample is a sample from a Normal population or not.

Example 5.6 (*Spar_Guthaben.sav*)
A question is the following: Are savings balances X (in €) Normally distributed? For $n = 12$ clients the following savings balances are observed:

9 800 9 300 15 200 8 600 12 200 11 600
10 200 8 700 6 900 9 600 15 500 7 200

We decompose the twelve observed values of the variable X into five equidistant classes. Each of the class has a width of 2 000.
Then we calculate the empirical density function= $\frac{\text{frequency of the class}}{2\,000}$:

class	frequency	density
6000 – 8000	2/12	$0.000\,08\overline{3}$
8000 – 10000	5/12	$0.000\,208\overline{3}$
10000 – 12000	2/12	$0.000\,08\overline{3}$
12000 – 14000	1/12	$0.000\,041\overline{6}$
14000 – 16000	2/12	$0.000\,08\overline{3}$

The unknown parameters μ and σ of the supposed Normal distribution are estimated as:

$\overline{x} = 10\,400$ is an estimator for μ
$s = \sqrt{s^2} = 2\,773.24883$ is an estimator for σ

We plot the histogram and the corresponding Normal curve $N(\mu = 10\,400; \sigma = 2\,773.24883)$ into one diagram:

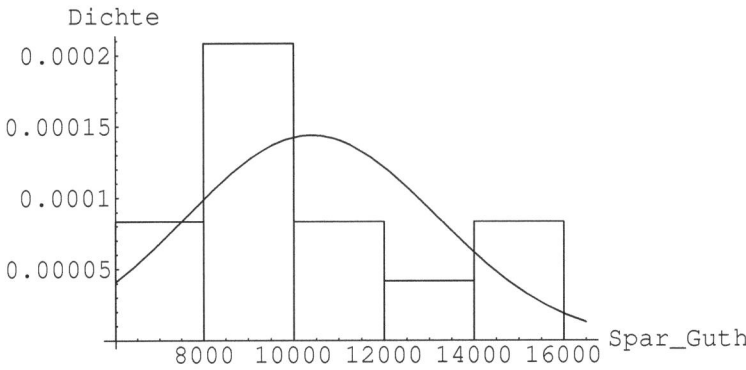

Histogram and Normal curve

The Normal curve (bell curve) is not a good fit for the histogram. Therefore, the Normal distribution is not a good fit for the amount of savings balance of a client.

If the histogram and the Normal curve fit well together then it can be assumed that the observed sample is a sample from a Normal distribution.

For "small" sample sizes n the histogram has the shape of a few rectangles, thus the fit of the bell curve with the histogram will never be good for n smaller than thirty.

5.5 Empirical Cumulative Distribution Function versus Normal Distribution

We plot the points of the empirical cumulative distribution function and the points of the corresponding Normal cumulative distribution function, brief: c.d.f., into one graph simultaneously. Then we check visually whether the observed sample is a sample from Normal distributions. We link the points of the empirical cumulative distribution function with lines for better interpretation. Linking the points we get a 45-degree line in our graphic box if there are no tied values in the sample observations. The graphic box with the 45-degree line and the points of the Normal distribution is called Plot-Point Diagram, brief: **P-P plot**.

The observed sample belongs to a Normal population if the points in the

P-P plot fit the 45-degree line quite well.

Example 5.7 (*Spar_Guthaben.sav*)
We want to know whether the random variable X = "savings balance (in €)
of a client" belongs to a Normal population or not. For $n = 12$ clients the
following savings balances were observed:

9 800 9 300 15 200 8 600 12 200 11 600

10 200 8 700 6 900 9 600 15 500 7 200

The unknown parameters μ and σ of the assumed Normal distribution are
estimated as:

$\bar{x} = 10 400$ is an estimator for μ
$s = \sqrt{s^2} = 2 773.24883$ is an estimator for σ

We get the following empirical cumulative distribution function $F\left(x_{(i)}\right)$ of
the ordered observed sample values $x_{(i)}$ and we get the following Normal
cumulative distribution function $F_U\left(z_{(i)}\right)$ of the ordered and standardized
sample values $z_{(i)}$:

i	$x_{(i)}$	$z_{(i)}$	$F(x_{(i)}) = \frac{i}{12}$	$F_U(z_{(i)})$
1	6900	−1.26	.0833	.1038
2	7200	−1.15	.1667	.1251
3	8600	−0.65	.2500	.2578
4	8700	−0.61	.3333	.2709
5	9300	−0.40	.4167	.3446
6	9600	−0.29	.5000	.3858
7	9800	−0.22	.5833	.4129
8	10200	−0.07	.6667	.4721
9	11600	0.43	.7500	.6664
10	12200	0.65	.8333	.7422
11	15200	1.73	.9167	.9582
12	15500	1.84	1.0000	.9671

We plot the 45-degree line and the points $F_U(z_{(i)})$ together into a graphic box to get the P-P plot:

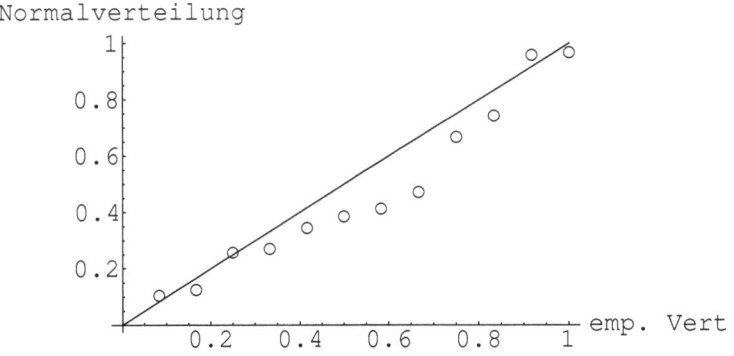

empirical cumulative distribution function and Normal c.d.f.

The points fit the 45-degree line quite well. Thus we can say that savings balances belong to a Normal population.

5.6 Quantile-Quantile Plot

If there are no ties (identical values) in the ordered sample $x_{(1)}, x_{(2)}, \ldots, x_{(n)}$, the empirical cumulative distribution function is $\frac{1}{n}, \frac{2}{n}, \ldots, \frac{n}{n} = 1$. For a visual check to see whether the sample belongs to a Normal distribution, we calculate the $\frac{i}{n}$-quantiles of the Normal distribution. The plot of the pairs $(x_{(i)}, \left(\frac{i}{n}\right)$-quantile) is called **QQ-Plot**.

Example 5.8 (*Spar_Guthaben.sav*)
X=Sparguthaben.
The ordered twelve observed values are:

i	$x_{(i)}$	$F(x_{(i)}) = \frac{i}{12}$	$\frac{i}{12}$-Quantil
1	6900	,0833	6564,613
2	7200	,1667	7717,099
3	8600	,2500	8529,472
4	8700	,3333	9205,486
5	9300	,4167	9816,430
6	9600	,5000	10400,000
7	9800	,5833	10983,570
8	10200	,6667	11594,514
9	11600	,7500	12270,528
10	12200	,8333	13082,901
11	15200	,9167	14235,387
12	15500	1,0000	∞

The pairs $(6\,900; 6\,564{,}613), (7\,200; 7\,717{,}099), \ldots, (15\,200; 14\,235{,}387)$ are plotted in a diagram (QQ-Plot). As a reference line we plot the 45-degree line, too:

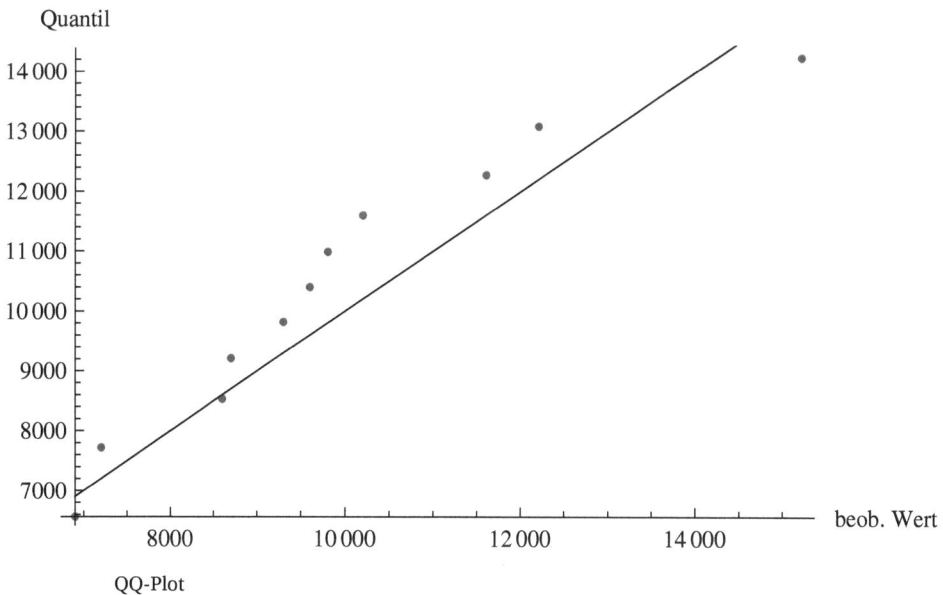

QQ-Plot

A normal distribution is assumed, if the points are plotted close to the 45-degree line.

5.7 Summary

Which procedure should be used to ensure Normal distribution? Firstly, a test should be preferred to a visual decision:

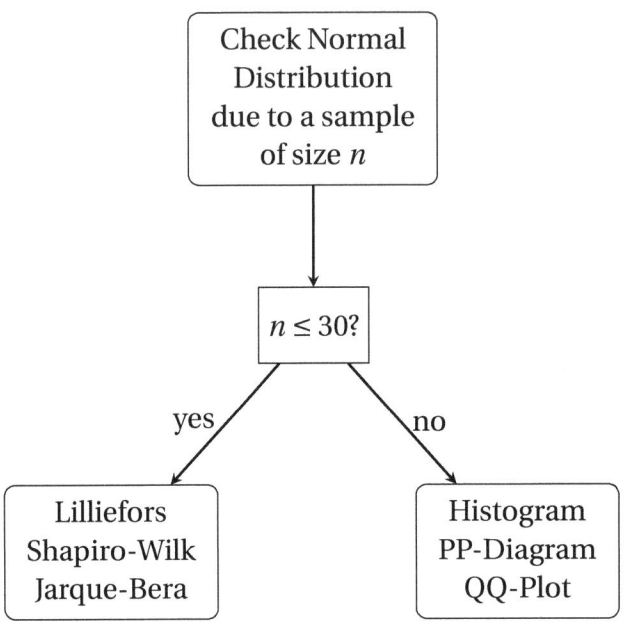

But which of the goodness-of-fit tests is the "best"? This means that if there is no Normal distribution, which test will approve the non Normal distribution most frequently?

In 1965 the authors S. S. Shapiro and M. B. Wilk presented in the journal Biometrika, Vol. 52, No. 3/4. (Dec., 1965), pp. 591-611 in the paper "An Analysis of Variance Test for Normality (Complete Samples)" the Shapiro-Wilk test. (cf. http://sci2s.ugr.es/keel/pdf/algorithm/articulo/shapiro1965.pdf) It pointed out (cf. page 608), that most frequently as its competitors the Shapiro-Wilk test detected, that there was no Normal distribution

Thorsten Thadewald and Herbert Büning conferred in their paper "Jarque-Bera Test and its Competitors for Testing Normality - A Power Comparison" in the Journal of Applied Statistics 2007, Volume 34, issue 1, pages 87-105 the Shapiro-Wilk test with the Jarque-Bera test. They got the following results:

"It turns out that for the Jarque-Bera test the approximation of critical values by the chi-square distribution does not work as well. The test is superior in power to its competitors for symmetric distributions with medium up to long tails and for slightly skewed distributions with long tails. The power of the Jarque-Bera test is poor for distributions with short tails, especially if the shape is bimodal, sometimes the test is even biased. In this case a modification of the Cramér-von Mises test or the Shapiro-Wilk test may be recommended."

Result: For symmetric distributions with a kurtosis of at least three, the Jarque-Bera test performs better than the Shapiro-Wilk test; for some distributions even the Kolmogorov-Smirnov test performs better than the Jarque-Bera test. But if there is a distribution with a kurtosis less than three, the Shapiro-Wilk test should be preferred:

Kurtosis	Skewness		
	$S < 0$	$S = 0$	$S > 0$
$K < 3$		Shapiro-Wilk	Shapiro-Wilk
$K = 3$		Jarque-Bera Shapiro-Wilk Lilliefors KS	
$K > 3$		Jarque-Bera	Jarque-Bera

For the sample size n we get the values S and K for the Jarque Bera test from SPSS as follows:

$$S = \frac{n-2}{\sqrt{n(n-1)}} \cdot S_{SPSS}$$

$$K = \frac{(n-2)(n-3)}{n^2-1} \cdot K_{SPSS} + 3 \cdot \frac{n-1}{n+1}$$

5.8   SPSS Commands

Input

We want to check whether the sample in example 5.2 is a sample from a Normal distribution. We type:

No	Sav._Balanc.
1	9 800
2	9 300
3	15 200
4	8 600
5	12 200
6	11 600
7	10 200
8	8 700
9	6 900
10	9 600
11	15 500
12	7 200

5.8.1   Lilliefors Test

Commands

1) Analyze → Descriptive Statistics → Explore ...

2) Dependent List = Savings_Balance
   (Factor List = Group)
   Select "Both" as "Display".

3) Click "Statistics".
   Select "Descriptives".
   Click "Continue".

4) Click "Plots".
   Select "Histogram" in "Descriptive".

Select "Normality plots with tests".
Click "Continue".

5) Click "OK".
The value 0.200 is the lower bound of the *p*-value.

Output

<table>
<tr><th colspan="4" style="text-align:center">Statistic</th></tr>
<tr><td></td><td></td><td>Statistic</td><td>Std. Error</td></tr>
<tr><td>Spar_Guth</td><td>Mean</td><td>10400.0000</td><td>800.56798</td></tr>
<tr><td>95% Confidence Interval</td><td>Lower Bound</td><td>8637.9618</td><td></td></tr>
<tr><td>for Mean</td><td>Upper Bound</td><td>12162.0382</td><td></td></tr>
<tr><td></td><td>5% Trimmed Mean</td><td>10311.1111</td><td></td></tr>
<tr><td></td><td>Median</td><td>9700.0000</td><td></td></tr>
<tr><td></td><td>Variance</td><td>7690909.091</td><td></td></tr>
<tr><td></td><td>Std. Deviation</td><td>2773.24883</td><td></td></tr>
<tr><td></td><td>Minimum</td><td>6900.00</td><td></td></tr>
<tr><td></td><td>Maximum</td><td>15500.00</td><td></td></tr>
<tr><td></td><td>Range</td><td>8600.00</td><td></td></tr>
<tr><td></td><td>Interquartile Range</td><td>3425.00</td><td></td></tr>
<tr><td></td><td>Skewness</td><td>.830</td><td>.637</td></tr>
<tr><td></td><td>Kurtosis</td><td>−.049</td><td>1.232</td></tr>
</table>

Tests of Normality

	Kolmogorov-Smirnov[a]			Shapiro-Wilk		
	Statistic	df	Sig.	Statistic	df	Sig.
Spar_Guth	0.195	12	.200*	.908	12	.199

a. Lilliefors Significance Correction

∗ This is a lower bound of the true significance.

The value 0.200 is the lower bound of the *p*-value; so the *p*-value is larger than 0.05; this means that we assume Normal distribution of the savings balance.

5.8.2 Shapiro-Wilk Test

Commands

1) Analyze → Descriptive Statistics → Explore ...

2) Dependent List = Savings_Balance
(Factor List = Group)
Select "Both" as "Display".

3) Click "Statistics".
   Select "Descriptives".
   Click "Continue".

4) Click "Plots".
   Select "Histogram" in "Descriptive".
   Select "Normality plots with tests".
   Click "Continue".

5) Click "OK".
   The *p*-value is 0.199.

Output

Statistic

			Statistic	Std. Error
Spar_Guth		Mean	10400.0000	800.56798
	95% Confidence Interval for Mean	Lower Bound	8637.9618	
		Upper Bound	12162.0382	
		5% Trimmed Mean	10311.1111	
		Median	9700.0000	
		Variance	7690909.091	
		Std. Deviation	2773.24883	
		Minimum	6900.00	
		Maximum	15500.00	
		Range	8600.00	
		Interquartile Range	3425.00	
		Skewness	.830	.637
		Kurtosis	−.049	1.232

Tests of Normality

	Kolmogorov-Smirnov[a]			Shapiro-Wilk		
	Statistic	df	Sig.	Statistic	df	Sig.
Spar_Guth	0.195	12	.200*	.908	12	.199

a. Lilliefors Significance Correction

∗ This is a lower bound of the true significance.

The *p*-value 0.199 is larger than 0.05; this means thta we assume Normal distribution of the savings balance.

5.8.3   Histogram versus Normal Distribution

We compare the histogram of the observed values with the corresponding Normal curve in one graphic box to check the assumption of Normality.

The shape of the histogram should approximately follow the shape of the Normal curve.

Commands

1) Analyze → Descriptive Statistics → Frequencies

2) Variable(s)=Spar_Guth

3) Click "Charts" and select "Histogram" and select "With Normal Curve". Click "Continue".

4) Click "OK".
   We get a diagram with the histogram and the corresponding Normal curve.

On the Y-axis in the SPSS diagram the absolute numbers (instead of the density) of cases in the associated class are labelled. But that makes no difference because we have equidistant classes.

5.8.4   Empirical c.d.f. versus Normal c.d.f.

The plotted points should follow the 45-degree line. Then the assumption of Normality is not violated.

Commands

1) Analyze → Descriptive Statistics → P−P Plots ...

2)   Variables =   Spar_Guth
   Test Distribution=   Normal

3) Click "OK".
   A P-P plot of Spar_Guth is displayed. The plotted points should follow the 45-degree line. Then the assumption of Normality is not violated.

5.8.5 QQ-Plot

The plotted points should follow the 45-degree line. Then the assumption of Normality is not violated.

Commands

1) Analyze → Descriptive Statistics → Q–Q Plots ...

2)   Variables =   Spar_Guth
      Test Distribution=   Normal

3) Click "OK".
   A QQ-Plot of Spar_Guth is displayed. The plotted points should follow the 45-degree line. Then the assumption of Normality is not violated.

6 Test of the Homogeneity of Two Variances

Main purpose: We want to check with a test whether the theoretical variances of two random variables are the same. The random variables must be leveled as follows:

nominal:	no
ordinal:	no
scale:	yes

Furthermore, the two variables whose variances are checked must be Normally distributed and stochastically independent.

6.1 Levene Test

If there are two random variables X and Y, it is sometimes of interest to know whether either equality or inequality of the two theoretical variances σ_x^2 resp. σ_y^2, can be assumed. If the two variances are unknown, we check with a test the equality:

Levene Test

H_0 : The two theoretical variances are the same;
     brief: $\sigma_x^2 = \sigma_y^2$

versus

H_1 : The two theoretical variances are not the same;
     brief: $\sigma_x^2 \neq \sigma_y^2$

Rejection of $H_0 \Leftrightarrow p$-value ≤ 0.05

The assumptions for the test are that X and Y are stochastically independent and each have a continuous distribution (not necessarily the Normal distribution).

On basis of two samples, one of each population, we want to test the null hypothesis of the Levene test.

Example 6.1 *Yieldings.sav)*

We want to invest money in stock. There are two alternative investments in stock. Let X denote the monthly yield (in percent) of the investment in stock I and let Y be the monthly yield (in percent) of the investment in stock II. (The German word of yield is "Rendite".) From the last months we get the following yields:

$$(x_1, \ldots, x_5) = (4.2; 3.7; 4.0; 3.8; 4.3)$$

$$(y_1, \ldots, y_6) = (4.0; 3.9; 4.0; 4.1; 4.0; 4.0)$$

The sample means are:

$$\bar{x} = \frac{1}{5}[4.2 + 3.7 + 4.0 + 3.8 + 4.3] = 4.0$$

$$\bar{y} = \frac{1}{6}[4.0 + 3.9 + 4.0 + 4.1 + 4.0 + 4.0] = 4.0$$

The sample variances and the sample standard deviations are:

$$s_x^2 = \frac{1}{4}\left[(4.2 - 4.0)^2 + \ldots + (4.3 - 4.0)^2\right] = \frac{0.05}{4} = 0.065$$

$$s_y^2 = \frac{1}{5}\left[(4.0 - 4.0)^2 + \ldots + (4.0 - 4.0)^2\right] = \frac{0.02}{5} = 0.004$$

$$s_x = \sqrt{0.065} = 0.254951$$

$$s_y = \sqrt{0.004} = 0.06324555$$

i.e. the sample variances differ.

We get the observed value of the Levene test statistic as:

$$F_{\text{emp.}} = \frac{N-k}{k-1} \cdot \frac{\sum_{i=1}^{k} n_i (\bar{z}_{i\cdot} - \bar{z}_{\cdot\cdot})}{\sum_{i=1}^{k} \sum_{j=1}^{n_i} (z_{ij} - \bar{z}_{i\cdot})^2} = 9.298$$

Notation:

$$
\begin{aligned}
k &= \text{2 (number of groups)} \\
n_i &= \text{sample size of the } i\text{-th Group} \\
N &= n_1 + n_2 + \ldots + n_k = 5 + 6 = 11 \\
z_{1j} &= |x_j - \bar{x}| = (0.2; 0.3; 0; 0.2; 0.3) \\
z_{2j} &= |y_j - \bar{y}| = (0; 0.1; 0, 0.1; 0; 0) \\
\bar{z}_{1\cdot} &= \tfrac{1}{n_1}[z_{11} + z_{12} + \ldots + z_{1n_1}] = \tfrac{1}{5}[0.2 + 0.3 + 0 + 0.2 + 0.3] = 0.2 \\
\bar{z}_{2\cdot} &= \tfrac{1}{n_2}[z_{21} + z_{22} + \ldots + z_{2n_2}] = \tfrac{1}{6}[0 + 0.1 + 0 + 0.1 + 0 + 0] = 0.\overline{3} \\
\bar{z}_{\cdot\cdot} &= \tfrac{1}{n_1 + n_2}[z_{11} + z_{12} + \ldots + z_{1n_1} + z_{21} + z_{22} + \ldots + z_{2n_2}] = 0.1\overline{09}
\end{aligned}
$$

We get the p-value of the Levene test from the F-distribution as follows:

$$P_{k-1;N-k}(F > F_{\text{emp.}}) = P_{1;9}(F > 9.298) = 0.01380938 \approx 0.014$$

Because the p-value is smaller than 0.05, you can assume that the two have significantly different variances. This means that we reject the null hypothesis; this means that the risks of the two investments differ.

If all theoretical variances are the same, we say we have **homogeneity** of the variances. If not, i.e. if at least two variances differ, we say we have **heterogeneity** of the variances.

Remark: The Levene test checks the null hypothesis of equality of two (or more) variances. If the variances are equal, the value of the empirical test statistic $F_{\text{emp.}}$ is close to zero or equals zero. If however the variances differ, the value of the empirical test statistic is much larger than zero. Is now found in comparison of two variances σ_x^2 and σ_y^2, that both variances differ significantly, the value of the empirical teststatistic is "large", no matter if $\sigma_x^2 > \sigma_y^2$ or $\sigma_x^2 < \sigma_y^2$. In particular the Levene test is inappropriate for one-sided testing problems.

6.2 Summary

The Levene test verifies whether two or more theoretical variances have the same value.

6.3 SPSS Commands

Input

We want to check whether the theoretical variances of the two yields in example 6.1 are the same. Type the eleven values into the SPSS Data View as follows:

No	Yield	Group
1	4.2	1
2	3.7	1
3	4.0	1
4	3.8	1
5	4.3	1
6	4.0	2
7	3.9	2
8	4.0	2
9	4.1	2
10	4.0	2
11	4.0	2

Commands

The samples must be descended from Normal distributions. The Shapiro-Wilk test has the two p-values 0.692 for stock I and 0.101 for stock II. Therefore, the yields of the two stocks can be assumed to be Normally distributed.

1) Open the file Yielding.sav

2) Analyze $\rightarrow$ Compare Means $\rightarrow$ Independent-Samples T Test ...

3) Test Variable = Yield
   Grouping Variable = Group
   Click "Define Groups".
   Type 1 as the Group 1 value and 2 as the Group 2 value.
   Click "Continue".

4) Click "OK".
   The column "Significance" in the "Levene's Test for Equality of Variances" table displays the p-value as 0.014.

Output

Independent Samples Test

		Levene's Test for the Equality of Variances		t-test for Equality of Means					95% Confidence Interval of the Difference	
		F	Sig.	t	df	Sig. (2-tailed)	Mean Difference	Std. Error Difference	Lower	Upper
Yield	Equal variances assumed	9.298	.014	.000	9	1.000	.00000	.10681	–.24161	.24161
	Equal variance not assumed			.000	4.411	1.000	.00000	.11690	–.31298	.31298

7 Test on Two Means

Main purpose: We want to check with a test whether the theoretical means of two random variables are the same. The random variables must be leveled as follows:

nominal:	no
ordinal:	no
scale:	yes

Furthermore, the two variables whose mean values are checked must be Normally distributed and stochastically independent.

In many situations it is necessary to compare two means when neither is known. Generally: We have two Normal populations - one with a random variable X that has a mean μ_X and variance σ_x^2, and other with a random variable Y that has a mean μ_Y and variance σ_y^2. The random variables X and Y must be stochastically independent.

If the two theoretical variances of X and Y are equal (homogeneous), the equality of the expected values is checked with the so called t-test. If the theoretical variances are unequal (heterogeneous), the equality of the expected values is checked with the Welch test.

We start with the case of same variances.

7.1 Two-sided t-Test

We assume that the two populations have the same variance.
When it can be assumed that the two populations have the same variance

(homogeneity of variances), the test is called **t-test** because the distribution of the test statistic is a t-distribution.

We illustrate an example.

Example 7.1 (*Birth_Smoking_35_41.sav* c.f. Daniel [2004] p. 464)
Smoking being pregnant is a risk for the unborn baby and the mother too. The doctor resp. woman doctor can identify smoking by the color of the placenta: The placenta is black instead of red.

Let X denote the weight at birth of a new born baby of a non smoking mother and let Y denote the weight at birth of a new born baby of a smoking mother.

The parameters are:

$\mu_x =$ expected weight (in g) at birth of a baby of a
      non smoking mother

$\mu_y =$ expected weight (in g) at birth of a baby of a
      smoking mother

We want to test the null hypothesis that the mean values of the weight at birth of babies of non smoking mothers and of babies of smoking mothers are identical:

Two-sided t-Test with independent samples
H_0: The expected weights at birth of a child of a non smoking mother and of smoking mother do not differ; brief: $\mu_x - \mu_y = 0$
versus
H_1: The expected weights at birth of a child of a non smoking mother and of smoking mother differ; brief: $\mu_x - \mu_y \neq 0$
Rejection of H_0 $\Leftrightarrow$ p-value ≤ 0.05

On basis of two samples, one of each population, we want to test the null hypothesis.

We have the weights at birth of 28 ($n_x = 16$, $n_y = 12$) new born babies and the smoker status of the corresponding mother, the birth happened between the 35th and the 41th week of pregnancy:

No	Weight	Status	No	Weight	Status
1	3330	N	15	2619	N
2	3450	N	16	2841	N
3	3130	N	17	2940	R
4	3226	N	18	2420	R
5	2729	N	19	2760	R
6	3410	N	20	2440	R
7	3095	N	21	2715	R
8	3244	N	22	3130	R
9	2520	N	23	2928	R
10	3523	N	24	3446	R
11	2920	N	25	2957	R
12	3040	N	26	2580	R
13	3322	N	27	3175	R
14	3459	N	28	2740	R

One of the assumptions of the t-test is the Normal distribution. We check the Normal distribution of the variable X="Weight at birth of a new born baby of a non smoking mother" and the Normal distribution of the variable Y="weight at birth of a new born baby of a smoking mother". The skewness for the weight of kids of a non smoking mother is $S = -.569$ (i.e. no test is recommended) and the p-value of the Shapiro-Wilk test is 0.317. The skewness for the weight of kids of a smoking mother is $S = 0.342$ and the kurtosis is $K = -0.234$; the p-value of the Shapiro-Wilk tests is 0.867. Hence both variables are Normally distributed.

The sample means and the sample deviations are:

$$\bar{x} = 3116.1250$$
$$\bar{y} = 2852.5833$$
$$s_x = 313.09783$$
$$s_y = 305.86969$$
$$\bar{x} - \bar{y} = 263.5417$$

I.e. on average, non-smoking children are 264 g heavier.

The p-value of the Levene test is 0.845; that means we have homogeneity of the two theoretical variances: $\sigma_x^2 = \sigma_y^2$ and therefore $\sigma_x = \sigma_y$. Furthermore we assume that the variables X and Y are stochastically independent. From the t distribution we get p-value as follows:

$$2 \cdot P_{n_x+n_y-2}\left(t > \left|\frac{\overline{x}-\overline{y}}{\sqrt{(\frac{n_x-1}{n_x+n_y-2}s_x^2 + \frac{n_y-1}{n_x+n_y-2}s_y^2)(1/n_x+1/n_y)}}\right|\right)$$

$$= 2 \cdot P_{26}(t > 2.225742) = 2 \cdot 0.0175 = 0.035$$

i.e. p-value ≤ 0.05; i.e. we reject the null hypothesis; i.e. the expected weights at birth in the two groups differ significantly.

Comment: In the sample the kids of a smoking mother weigh 263.5417 g less on average than the kids of a non smoking mother. That difference is a statement only for the sample. The difference in the population is unknown. However, the difference of 263.5417 g in the sample suffices that the test can claim that in the population the mean weight of babies of a non smoking mother and of babies of a smoking mother differ significantly. Provided that the samples are representative.

Pediatrics Vol. 111, page 1 318: *Tobacco-exposed infants were more excitable and hypertonic, required more handling and showed more stress/abstinence signs, specifically in the central nervous system (CNS), gastrointestinal, and visual areas. Dose-response relationships showed higher maternal salivary cotinine values related to more stress/abstinence signs including CNS and visual stress and higher excitability scores. Cigarettes per day during pregnancy was related to more stress/abstinence signs including CNS and visual stress.*

7.2 Two-sided Welch Test

If the Levene-Test rejects based on the sample variances s_x^2 and s_y^2 the null hypothesis of equal theoretical variances σ_x^2 and σ_y^2, we have heterogeneity of the variances and SPSS runs the so called **Welch test**.

To run a Welch test the following assumptions must be fulfilled: Normal distribution of each variable, stochastic independence of the variables, heterogeneity of the variances of the variables.

Example 7.2 (*Internet_F_M.sav*)
We want to know whether women and men spend the same time working in the internet.

We consider the two random variables:

X = Time (in hours per week) surfing the Internet of a man

Y = Time (in hours per week) surfing the Internet of a woman

Two-sided Welch Test

H_0:   Women and men spend the same expected time surfing the Internet; brief: $\mu_x - \mu_y = 0$

versus

H_1:   Women and men spend a various expected time surfing the Internet; brief: $\mu_x - \mu_y \neq 0$

Rejection of $H_0 \Leftrightarrow p$-value $\leq \alpha$

We are doing a survey to make the test decision (hours per week):

No.	Time	Gender
1	3	f
2	3.5	f
3	4	f
4	2.5	f
5	3	f
6	4.5	f
7	2.5	f
8	3	f
9	8	m
10	3	m
11	10	m
12	4	m
13	16	m
14	5	m
15	14	m
16	7.5	m
17	9	m
18	8	m
19	22	m
20	4	m

One of the assumptions of the Welch test is that both variables X="Time (in hours per week) surfing the Internet of a man" and Y="Time (in hours per week) surfing the Internet of a woman" must have Normal distribution. The skewness is 0.808 and the kurtosis is -0.229 for women. The p-value of the Shapiro-Wilk test is 0.273. For men the skewness and the kurtosis are $S = 1.163$ resp. $K = 1.066$; the value of the Jarque-Bera test statistic is $T \approx \frac{12}{6}\left(1.163^2 + \frac{1.066^2}{4}\right) = 3.273 < 5.991$. The p-value of the Shapiro-Wilk-Test is 0.128 (male). Hence both variables are Normally distributed.

Furthermore we assume that the two variables are stochastically independent.

The p-value of the Levene-Test is 0.010; this means that the variances of X and Y differ significantly.

The sample means and the standard sample deviations are:

$$\bar{x} = 9.2083$$
$$\bar{y} = 3.25$$
$$s_x = 5.63858$$
$$s_y = 0.70711$$
$$\bar{x} - \bar{y} = 5.95833$$

This means that men spend about six hours more per week on average surfing the Internet than women do.

The test statistic of the Welch test has a t-distribution with $\dfrac{(1+R)^2}{\frac{R^2}{n_x-1} + \frac{1}{n_y-1}} =$

11.515 degrees of freedom, with $R = \dfrac{s_x^2 \cdot n_y}{s_y^2 \cdot n_x} = \dfrac{5.63858^2 \cdot 8}{0.70711^2 \cdot 12} = 42.39106$.

The p-value of the Welch test is:

$$p\text{-value} \quad = 2 \cdot P_{11.280}\left(t > \left| \frac{\bar{x} - \bar{y}}{\sqrt{s_x^2/n_x + s_y^2/n_y}} \right| \right)$$

$$= 2 \cdot P_{11.515}(t > | 3.618 |) = 0.003760569 \approx 0.004$$

This means that women and men are surfing significantly a different expected time the Internet.

Comment: In the sample, women surfed on average 5.95833 hours shorter the Internet than men. That difference is a statement only for the sample. The difference in the population is unknown. However, the difference of 5.95833 hours in the sample suffices that the test can claim that in the population the mean time of women and men surfing the Internet differs significantly. Provided that the samples are representative.

7.3 One-sided t-Test and One-sided Welch Test

The test that checks the equality of two means, is called two-sided test, because the alternative H_1 includes both options: That in example 7.1 babies of smoking mothers are lighter than babies of non-smoking mothers resp.

that babies of smoking mothers are heavier than babies of non-smoking mothers. If a difference was found, the question arises as to whether babies from smoking mothers are significantly lighter or significantly heavier than babies from non-smoking mothers. The test that reveals the sign of the difference $\mu_X - \mu_Y$, i.e. $\mu_X - \mu_Y > 0$ resp. $\mu_X - \mu_Y < 0$, is called one-sided test.

Example 7.3 (*Geburt_Raucherstatus_35_41.sav* c.f. Daniel [2004] p. 464) In the sample of example 7.1 the smoker babies are lighter than the non smoker babies. We want to know whether smoker babies are significantly (generally) lighter than non smoker babies in the population. A difference is called a significant difference, if the null hypothesis is rejected. Therefore the alternative H_1 is the statement: "Smoker babies are lighter than non smoker babies":

One-sided t-Test two independent samples:
H_0: Mean weight at birth of non smoker babies $\leq$ mean weight at birth of smoker babies brief: $\mu_X \leq \mu_Y$
versus
H_1: Mean weight at birth of non smoker babies $>$ mean weight at birth of smoker babies brief: $\mu_X > \mu_Y$
Rejection of $H_0 \Leftrightarrow p$-value (one-sided) ≤ 0.05

Considering the difference $X - Y$ we get the following one-sided testing problem:

H_0: Mean weight at birth of non smoker babies minus mean weight at birth of smoker babies ≤ 0
   brief: $\mu_X - \mu_Y \leq 0$

H_1: Mean weight at birth of non smoker babies minus mean weight at birth of smoker babies > 0
   brief: $\mu_X - \mu_Y > 0$

The null hypothesis H_0 will be rejected if and only if the p-value equals 0.05 or is smaller than 0.05.

The test is based on a sample.

The p-value is calculated due to students distribution (t-distribution):

$$P_{n_x+n_y-2}(t > t\text{emp.}) = P_{26}(t > 2.226) = 0.0175$$

This means p-value $\leq 0{,}05$; this means that we reject the null hypothesis; this means that the mean weight at birth of smoker babies is significantly smaller than the mean weight at birth of non smoker babies.

The p-value of the one-sided test is the p-value of the two-sided test divided by two:

$$p\text{-value (one-sided)} = \frac{p\text{-value (two-sided)}}{2} = \frac{0.035}{2} = 0.0175$$

Interpretation: In the sample smoker babies weighed on average 263.5417 g less than non smoker babies. This difference of 263.5417 g is a statement referring only to the sample. As it is in the population, it is unknown, but the difference of 263.5417 g in the sample claims that in the population the mean weight at birth of smoker babies is significantly smaller than the mean birth weight at birth of non smoker babies, if both samples were representative.

If a test shows that two means μ_X and μ_Y differ, the question immediately arises as to which sign the difference $\mu_X - \mu_Y$ has.

Example 7.4 (*workers.sav* c.f. Anderson et al. [2017] p. 354)
Do union workers and non-union workers have the same hourly wage rates?

In the file *workers.sav* we consider the hourly wage rate of $n_x = 15$ union workers and of $n_y = 20$ non-union workers:

Union Workers
22.40	18.90	16.70	14.05	16.20	20.00	16.10	16.30
19.10	16.50	18.50	19.80	17.00	14.30	17.20	

Non-Union Workers

17.60	14.40	16.60	15.00	17.65	15.00	17.55	13.30
11.20	15.90	19.20	11.85	16.65	15.20	15.30	17.00
15.10	14.30	13.90	14.50				

The two theoretical variances of the random variables X = "hourly wage rate of a union worker" and Y = "hourly wage rate of a non union worker" are homogeneous due to the p-value of the Levene-Tests 0.526.

The Normal distribution of the hourly wage rates is confirmed:

| Union | p-value | |
Worker	Lilliefors Test	Shapiro-Wilk Test
yes	≥ 0.2	0.651
no	≥ 0.2	0.860

We call μ_X = "mean value of the hourly wage rate of a union worker" and μ_Y = "mean value of the hourly wage rate of a non-union worker". We get the following two-sided t-Test:

$$H_0 : \mu_X - \mu_Y = 0 \text{ versus } H_1 : \mu_X - \mu_Y \neq 0$$
$$\text{Rejection of } H_0 \Leftrightarrow p\text{-value} \leq \alpha$$

Remark: This test is called "two-sided" test due to the alternative H_1, because H_1 includes the two sides of zero: smaller zero and larger zero. SPSS denotes the two-sided test as "two-tailed" test.

The p-value of the two sided t-Test is 0.005, i.e. there are significant differences among the expected rate of union workers and the expected rate of non-union workers.

Which group has the higher rate? We want to know the sign of the difference, is the difference $\mu_X - \mu_Y$ negative or positive?

In the sample we have observed 15 union workers with an average rate of 17.5367 and 20 non-union workers with an average rate of 15.36. Thus in the sample the union workers have the higher rate. Does this also apply in the population? To check this we run the one-sided t-test. The alternative of the one-sided t-test is like the statement in the sample: union workers have a higher rate than non-union workers. And the null hypothesis is the opposite statement: union workers don't have a higher rate than non-union workers.

One-sided t-test:

$$H_0 : \mu_X - \mu_Y \leq 0 \text{ versus } H_1 : \mu_X - \mu_Y > 0$$
$$\text{Rejection of } H_0 \Leftrightarrow p\text{-value (one-sided)} \leq \alpha$$

where p-value of the one-sided t-test $= \dfrac{p\text{-value of the two-sided } t\text{-test}}{2} =$
$\dfrac{0.005}{2} = 0.0025 \leq 0.05$

i.e. the mean hourly wage rate of union workers is significantly larger than the hourly wage rate of non-union workers; i.e. in the population union workers have significantly higher rates than non-union workers.

Remark: This test is called "one-sided" test due to the alternative H_1, because H_1 includes only one side of zero: larger zero.

7.4 Summary

The t-test verifies whether two expected values are the same if the two corresponding theoretical variances are the same. The Welch test verifies whether two expected values are the same if the two corresponding theoretical variances are not the same.

SPSS looks about the data in the sample for the formulation of the alternative hypothesis H_1 of a one-sided test:

Test value = 0		
Sample	Test	Testing Problem
	two-sided	$H_0: \mu_x - \mu_y = 0$ versus $H_1: \mu_x - \mu_y \neq 0$ p-value (two-sided)
$\bar{x} > \bar{y}$	one-sided	$H_0: \mu_x - \mu_y \leq 0$ versus $H_1: \mu_x - \mu_y > 0$ p-value (one-sided) $= 0.5 \cdot p$-value (two-sided)
$\bar{x} < \bar{y}$	one-sided	$H_0: \mu_x - \mu_y \geq 0$ versus $H_1: \mu_x - \mu_y < 0$ p-value (one-sided) $= 0.5 \cdot p$-value (two-sided)

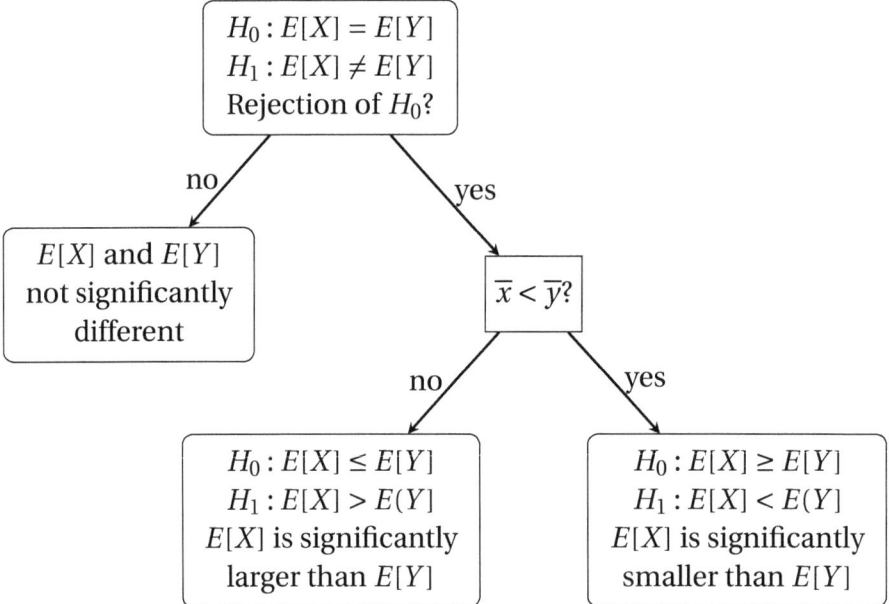

(If the assumption of the Normal distribution is not fulfilled see chapter 11.1.)

7.5 SPSS Commands

Input

We want to check with the sample of example 7.1, whether the mean weight of babies of non-smoking mothers and smoking mothers is the same. Type the data into the SPSS Data View as follows:

No	Weight	Group
1	3 330	1
2	3 450	1
3	3 130	1
⋮	⋮	⋮
16	2 841	1
17	2 940	2
⋮	⋮	⋮
28	2 740	2

The two samples must be samples from the Normal distribution otherwise, the sample sizes must be larger or equal 30 (rule of thumb for the Central-limit Theorem). Furthermore the variables X, Y must be stochastically independent.

Commands

1) Open the file "Birth_Smoking.sav"

2) Analyze → Compare Means → Independent-Samples T Test ...

3) Test Variable = Weight
   Grouping Variable = Group
   Click "Define Groups".
   Type 1 as the Group 1 value and 2 as the Group 2 value.
   Click "Continue".

4) Click "OK".
   The column "Significance (2-tailed)" in the output table "t-test for Equality of Means" displays the p-value of the two-sided test in the row "Equal variances assumed" as 0.035.

Output
⚠ SPSS refers to the Welch test as a t-Test. The selection, which of the two tests should be considered, is made due to the p-value of the Levene-Test.

If the variances are homogeneous, as in example 7.1, the p-value of the t-Test is listed in the row "Equal variances assumed". If the variances are heterogeneous, as in example 7.2, the p-value of the Welch test is listed in the row "Equal variances not assumed". In particular, the SPSS commands for the t-test and the Welch test are identical.

Output Birth Weights

Independent Samples Test

		Levene's Test for Equality of Variances		t-test for Equality of Means					95% Confidence Interval of the Difference	
		F	Sig.	t	df	Sig. (2-tailed)	Mean Difference	Std. Error Difference	Lower	Upper
Weight	Equal Variances assumed	.039	.845	2.226	26	.035	263.54167	118.40625	20.15414	506.92919
	Equal Variances not assumed			2.233	24.147	.035	263.54167	117.99681	20.08645	506.99689

Output Time in Internet

Independent Samples Test

| | | Levene's Test for Equality of Variances | | t-Test for Equality of Means | | | | | | 95% Confidence Interval of the Difference | |
		F	Sig.	t	df	Sig. (2-tailed)	Mean Difference	Std. Error Difference		Lower	Upper
Time	Equal Variances assumed	8.165	.010	−2.947	18	.009	−5.95833	2.02195		−10.20630	−1.71036
	Equal Variances not assumed			−3.618	11.515	.004	−5.95833	1.64680		−9.56324	−2.35342

8 Regression Analysis

Main purpose: Linear regression is used to model the value of a variable Y based on its linear relationship to one or more variables $X_1, X_2, \ldots, X_p$.

In the linear regression model:

$$Y \approx b_0 + b_1 \cdot x_1 + b_2 \cdot x_2 + \ldots b_p \cdot x_p$$

The aim is to determine this linear combination in order to obtain forecasts for the values of Y.
The variable Y depends on the values of the predictors $X_1, X_2, \ldots, X_p$. Therefore the variable Y is called dependent variable, the variables $X_1, X_2, \ldots, X_p$ are called independent variables.
The values $b_1, b_2, \ldots, b_p$ are called **regression coefficients** and the value b_0 is called constant.

⚠ In statistics there are two kinds of "independent" variables. In a linear model the value of the variable Y is explained by a few independent variables. And variables may be stochastically independent. Please don't switch these two terms of independence.

8.1 Multiple Linear Regression

The dependent variable Y must be leveled as follows:
nominal:	no
ordinal:	no
scale:	yes

The independent variables $X_1, X_2, \ldots$ must be leveled as follows:

nominal: no
ordinal: no
scale: yes

The data set from the example 8.1 will be used to get to know the regression analysis. We start with one independent variable in chapter 8.1.1 and two or more independent variables in chapter 8.1.2.

Example 8.1 (*Miles_Per_Gallon.sav* c.f. Berenson et al. [2015] p. 671)
A consumer organization wants to predict gasoline mileage (as measured by miles per gallon) based on horsepower of the car's engine and the weight of the car (in pounds).

Y = MPG (miles per gallon)
X_1 = weight (in pounds)
X_2 = horsepower (PS)

The aim of a regression analysis is always to include as few independent variables as possible in the model. As we will see later, the correlation between miles per gallon and curb weight is stronger than the correlation between miles per gallon and horsepower. Therefore, the independent variable X_1="Weight" (and not X_2="Horsepower") is first included in the model.

A sample of 50 recent car models was selected with the results as follows:

No.	MPG	PS	Weight	No.	MPG	PS	Weight
1	43.1	48	1985	26	23.9	90	3420
2	19.9	110	3365	27	29.9	65	2380
3	19.2	105	3535	28	30.4	67	3250
4	17.7	165	3445	29	36.0	74	1980
5	18.1	139	3205	30	22.6	110	2800
6	20.3	103	2830	31	36.4	67	2950
7	21.5	115	3245	32	27.5	95	2560
8	16.9	155	4360	33	33.7	75	2210
9	15.5	142	4054	34	44.6	67	1850
10	18.5	150	3940	35	32.9	100	2615
11	27.2	71	3190	36	38.0	67	1965
12	41.5	76	2144	37	24.2	120	2930
13	46.6	65	2110	38	38.1	60	1968
14	23.7	100	2420	39	39.4	70	2070
15	27.2	84	2490	40	25.4	116	2900
16	39.1	58	1755	41	31.3	75	2542
17	28.0	88	2605	42	34.1	68	1985
18	24.0	92	2865	43	34.0	88	2395
19	20.2	139	3570	44	31.0	82	2720
20	20.5	95	3155	45	27.4	80	2670
21	28.0	90	2678	46	22.3	88	2890
22	34.7	63	2215	47	28.0	79	2625
23	36.1	66	1800	48	17.6	85	3465
24	35.7	80	1915	49	34.4	65	3465
25	20.2	85	2965	50	20.6	105	3380

Miles_Per_Gallon.sav

We want to use a linear regression model $Y \approx b_0 + b_1 \cdot x_1 + b_2 \cdot x_2$ for the data set to predict the value of the dependent variable Y=gasoline mileage. At first we will consider only one independent variable.

8.1.1 One Independent Variable

A linear model $Y \approx b_0 + b_1 \cdot x_1$ with one independent variable is called simple regression model.

Example 8.2 (*Miles_Per_Gallon.sav* c.f. Berenson et al. [2015] p. 671)
We want to calculate the gasoline mileage Y based on the weight x_1. To check whether the relationship is a linear relationship we construct the scatter plot between the weight on the x-axis and the gasoline mileage on the y-axis:

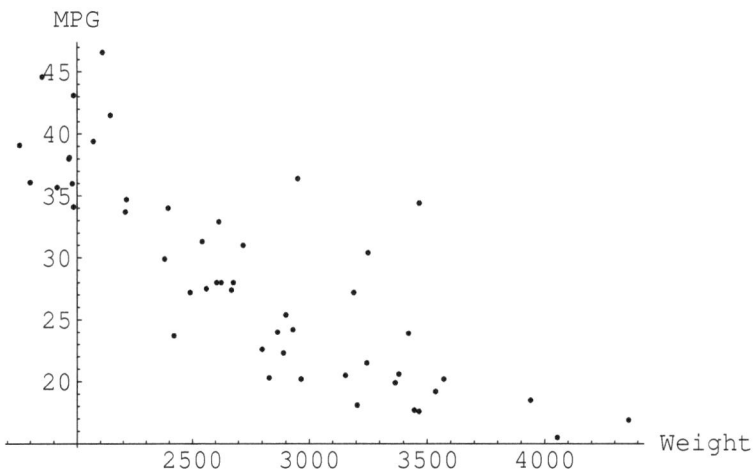

The points in the scatter plot are lying approximately on a descending line.

Linear Relationship

A simple linear model $Y \approx b_0 + b_1 \cdot x_1$ assumes a linear relationship between the y-values and the x-values. This assumption must be checked based on the bivariate data set.

The value of the Bravais-Pearson correlation coefficient indicates whether there is a linear relationship between the dependent variable Y and the independent variable X_1 or not.

Example 8.3 (*Miles_Per_Gallon.sav* c.f. Berenson et al. [2015] p. 671)
The correlation between the gasoline mileage and the weight is -0.825, i.e. negative strong correlation. The gasoline mileage depends on the weight very well. The negative sign of the correlation indicates that the heavier the car, the lower the mileage.

A strong correlation between Y and X_1 indicates, that there is a linear relationship between the two variables. A moderate or weak correlation indicates that we have to look for a non-linear relationship between the two variables.

Example 8.4 (*Miles_Per_Gallon.sav* c.f. Berenson et al. [2015] p. 671)
The correlation r(weight, horsepower) = 0.742 between weight and horsepower is a spurious correlation: If we only consider cars with the same gasoline mileage we would obtain the correlation r(weight, horsepower) = 0.264 between weight and horsepower. In order not to have to start a survey again, in which only cars with the same mileage are considered, the influence of the mileage could be stopped if we consider the so called **partial correlation** with the gasoline mileage as a control variable.

In this context SPSS calls a partial correlation a correlation of first order. And a correlation of Bravais- Pearson a correlation of zero order.

The correlation of Bravais-Pearson between MPG and weight is -0.825. If we only consider cars with the same horsepower we would get a correlation r(MPG, Weight)=-0.582 between MPG and weight. To get this correlation we calculate the partial correlation with horsepower as a control variable.

The correlation r(MPG, Horsepower) = -0.788 between mileage and horsepower is a negative moderate correlation. If we only consider cars with the same curb weight we would get the correlation r(MPG, horsewpower) = -0.465 between MPG and horsepower. To get this correlation we calculate the partial correlation with curb weight as a control variable.

Forecasting
In case of a linear relationship between the two variables Y and X_1 a line may describe the relationship between Y and X_1.

Example 8.5 (*Miles_Per_Gallon.sav* c.f. Berenson et al. [2015] p. 671)
We consider the simple linear regression model:

$$\boxed{\text{MPG}} \approx b_0 + b_1 \cdot \boxed{\text{Weight}}$$

respectively:
$$Y \approx b_0 + b_1 \cdot x_1$$

The values b_0 and b_1 cannot be calculated, they can only be estimated. We will estimate the values of b_0 and b_1 with the least squares method and designate the estimated values with $\widehat{b}_0$ and $\widehat{b}_1$. (It is common in statistics to use a hat over a parameter name for estimated parameter values.) The least squares method plots a line that minimizes the sum of the squares of the deviations between the observed values y_i of the dependent variable Y and the values $\widehat{b}_0 + \widehat{b}_1 \cdot x_i$ of Y on the line:

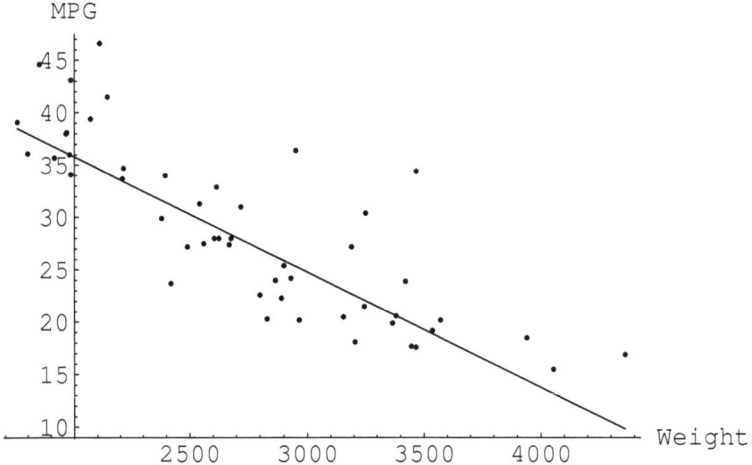

The least squares method provides:

$$\widehat{b}_0 = 57.797$$
$$\widehat{b}_1 = -0.011$$

The data set can be explained by a descending line. As the weight increases by one pound, the mileage decreases by 0.011 miles.
The line $y = 57.797 - 0.011 \cdot x$ is called **regression line**.
Interpretation of the estimated regression coefficient $\widehat{b}_1$: the value -0.011 is an estimated value of the expected decrease in gasoline mileage corresponding to an increase of one pound in the weight.

In a simple linear regression model (that is one independent variable), we comment the estimated regression coefficient $\hat{b}_1$ as an estimated value of the change in y for a one-unit change in the independent variable.

We use the regression line to predict values of Y.

With a weight of 2 250 pounds we get a gasoline mileage as follows:

$$57.797 - 0.011 \cdot 2250 = 33.9$$

This means that we must expect about 34 miles per gallon of a car with a weight of 2 250 pounds.

With a weight of 4 500 pounds we get a gasoline mileage as follows:

$$57.797 - 0.011 \cdot 4500 = 10.03825$$

This means that we must expect about 10 miles per gallon of a car with a weight of 4 500 pounds.

Are the values 34 resp. 10 miles good or bad predicted values?

We want to know whether a predicted value is reliable. For the answer we have to check whether the value is an interpolated or extrapolated value and we have to calculate the strength of correlation.

Because the weight 4 500 is not lying between the 1 755 = minimum and 4 360 = maximum of the observed values we call the predicted value of 10.0 miles an **extrapolated value**. Extrapolated predicted values are never reliable. Therefore the predicted 10.0 miles of a car of the weight of 4 500 pounds is not reliable.

Whereas the predicted value 33.9 miles is an **interpolated value** because

2 250 is lying in the interval [1 755;4 360].

The correlation between the gasoline mileage and the weight is −0.825, i.e. negative strong correlation. The gasoline mileage depends on the weight very well.

If a predicted value is an interpolated value and in addition if the correlation is strong, the predicted value is reliable. Therefore the predicted 33.9 miles of a car of a weight of 2 250 pounds is reliable.

Result: A predicted value is reliable if two conditions are fulfilled:

Reliability of forecast

Type of forecast	Correlation		
	weak	moderate	strong
Interpolation	no	no	yes
Extrapolation	no	no	no

Heteroscedasticity

There are several methods to estimate the constant b_0 and regression coefficient b_1 in the model $Y \approx b_0 + b_1 \cdot x_1$. One of these methods is the least squares method. The least squares method is a good method, if the points in the scatter plot are scattering in the same way around the regression line. The least squares method is inappropriate, if the points in the scatter plot do not scatter in the same way around the regression line.

A measure for the scattering is the **deviation**.

Example 8.6

The sample deviation of the data set 2,2,3,1,2, is smaller than the sample deviation of the data set 2,3,1,3,1; this means that the points in the first data set are lying closer to $\bar{x} = 2$ than the points in the second data set.

In a simple linear model $Y \approx b_0 + b_1 \cdot x_1$ the values y_1, y_2, ..., y_n of Y are called the observed values of Y. In contrast the n values of $\widehat{b_0} + \widehat{b_1} \cdot x_1$ are called the predicted values of Y. The n differences $y_i - b_0 - b_1 \cdot x_1$ (observed value minus predicted value) are called the **residuals**. The residuals are the distances of the points from the regression line.

In the example 8.5 the first observed value of Y=MPG is 43.1 miles. If we use the least squares method the first predicted value is $57.797 - 0.011 \cdot 1985 = 36.73023$ miles, therefore the first residual is $43.1 - 36.73023 = 6.36977$ miles. The second observed value of Y=MPG is 19.9 miles. If we use the least squares method the second predicted value is $57.797 - 0.011 \cdot 3365 = 22.08413$ miles, therefore the second residual is $19.9 - 22.08413 = -2.18413$ miles.

In a simple linear model $Y \approx b_0 + b_1 \cdot x_1$ the estimators of b_0 and b_1 based on the bivariate sample of size n. The least squares method assumes, that the residuals $Y - (b_0 + b_1 \cdot x_1)$ are random variables for all n values of X_1 with same theoretical variances $\sigma_1^2 = \sigma^2, \ldots, \sigma_n^2 = \sigma^2$. The least squares method is not valid (i.e. it is a wrong method) if the variances are not the same. The case of unequal variances is called **heteroscedasticity**. The case of equal variances is called **homoscedasticity**. In case of heteroscedasticity the values of b_0 and b_1 estimated by the least squares method are not precise, i.e. are incorrect.

Example 8.7 (*Miles_Per_Gallon.sav* c.f. Berenson et al. [2015] p. 671)
We consider the scatter plot with the regression line of the model MPG $\approx b_0 + b_1 \cdot$ Weight:

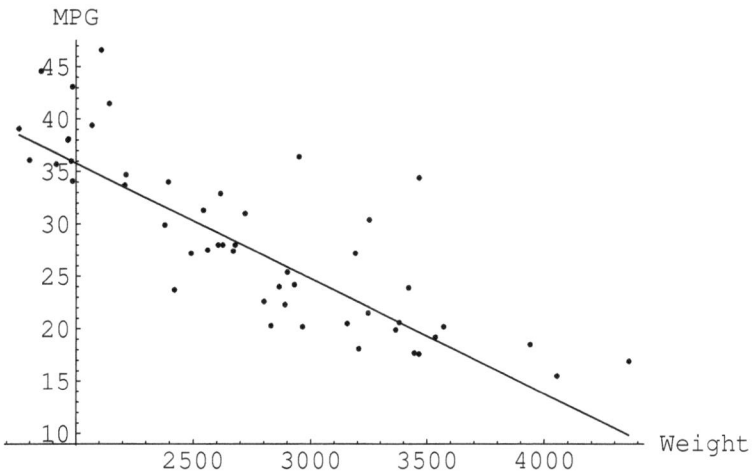

The points are scattering in the same way around the regression line, i.e. we get homoscedasticity.

The scatter plot between y and x_1 gives an overall impression whether we have hetero- or homoscedasticity.

Example 8.8
Let us consider the numeric values:

x_1	y
1	2
2	2
3	9
4	4
5	15
6	6
7	21
8	8
9	27
10	10

With the least squares method we get the following regression line:

$$y = 0.933 + 1.721 \cdot x_1$$

Now we construct the scatter plot between x_1 and y:

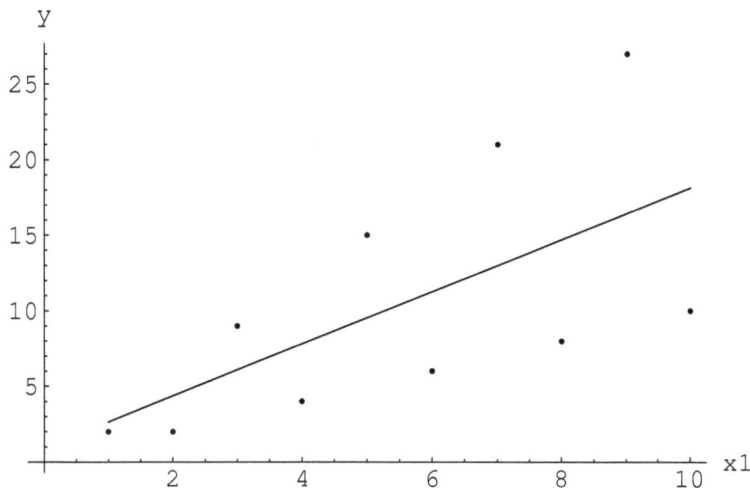

We get a v-formation. So we have heteroscedasticity. This means that we should not use the least squares method to estimate the constant and the regression coefficient.

High Leverage Points

Every observation (x_i, y_i) of a bivariate data set $(x_1, y_1), \ldots, (x_n, y_n)$ has an influence on the slope of the regression line. But there are high influential points and low influential points. We can see this, if we change the y-value in the scatter diagram. For every observation (x_i, y_i) SPSS computes the influence of this point on the slope of the regression line. The computed values are called **leverage values**. The leverage value of an observation (x_i, y_i) is determined by how far the value x_i of the independent variable is from its mean value. An influential observation may be an outlier (an observation with a y-value that deviates substantially from the "trend"). Outliers may have a dramatic effect on the slope of the regression line and must be removed from the data set.

Example 8.9

Consider the data set $(x_1, y_1), \ldots, (x_4, y_4) = (1, 2), (2, 4), (3, 5), (4, 1)$. We insert the values in the following applet:

1) Mozilla Firefox

2) *http://www.shodor.org/interactivate/activities/Regression/*

3) Plot points. We insert the four points in the diagram.

4) Display line of best fit.

5) Move points. If we move at $x_4 = 4$ the value y-value $y_4 = 1$ down or up, e.g. $y_4 = 5$, the slope of the regression line will change dramatically; this means that (x_4, y_4) is a high influential point.

6) Whereas (x_2, y_2) and (x_3, y_3) are low influential points.

To get the leverage values of every data point in SPSS we have to select "Save". Under "Distances" select "Leverage values". SPSS calculates for every data point the associated leverage value. The $n = 4$ leverage values are all lying between 0 and $\frac{n-1}{n} = \frac{4-1}{4} = 0.75$. Larger leverage values indicate a higher influence on the slope of the regression line.

SPSS shows the leverage values in the column LEV of the Data View as follows:

	x	y	LEV_1
1	1	2	0.45
2	2	4	0.05
3	3	5	0.05
4	4	1	0.45

This means that (x_1, y_1) and (x_4, y_4) are high influential points whereas (x_1, y_1) and (x_4, y_4) are low influential points.

8.1.2   Two Or More Independent Variables

A linear model with two or more independent variables is called multiple regression model.

Example 8.10 (*Miles_Per_Gallon.sav* see Berenson et al. [2015] p. 671)
We want to calculate the gasoline mileage Y based on the weight x_1 and on the horsepower x_2. This means that we consider the following linear model:

$$\text{Model: } Y \approx b_0 + b_1 \cdot x_1 + b_2 \cdot x_2 \text{ (two independent variables)}$$

8.1.3 Multiple Correlation Coefficient

To check whether a linear regression model with more than one independent variable makes sense, we consider the so-called **multiple correlation coefficient** R. The values of R are lying in the interval $[0;1]$. The interpretation is as follows: Values of R in the interval $(0;0.5)$ indicate a weak correlation. Values of R in the interval $(0.5;0.8)$ indicate a moderate correlation. Values of R in the interval $(0.8;1)$ indicate a strong correlation.

Example 8.11 (*Miles_Per_Gallon.sav* c.f. Berenson et al. [2015] p. 671)
The multiple correlation between gasoline mileage, weight and horsepower is 0.866, i.e. strong correlation.

The squared value $R^2 = 0.749$ is called **coefficient of determination**.

About 74.9% of the total variance is explained by the variance of the values $\widehat{b}_0 + \widehat{b}_1 x_1 + \widehat{b}_2 x_2$. This means that there is a good fit between the multiple linear regression model and the data set.

The differences $y_i - \widehat{y}_i$ of the observed values y_i minus the predicted values $\widehat{y}_i$ are called residuals. The most important indicator of the goodness of fit of a linear model is the sum of the squared residuals $\sum(y_i - \widehat{y}_i)^2$. The coefficient of determination R^2 is based on the follwing sum:

$$R^2 = 1 - \frac{\sum_{i=1}^{n}(y_i - \widehat{y}_i)^2}{\sum_{i=1}^{n}(y_i - \overline{y})^2}$$

If we have a perfect linear relationship, the value of the sum of the squared residuals is zero and we get the following value of the coefficient of determination:

$$R^2 = 1 - \frac{\sum_{i=1}^{n}(y_i - \widehat{y}_i)^2}{\sum_{i=1}^{n}(y_i - \overline{y})^2} = 1 - \frac{0}{s_y^2} = 1$$

If we have no linear relationship, the sum of the squared residuals may have the maximum value $\sum_{i=1}^{n}(y_i - \overline{y})^2$ and we get the following value of the coefficient of determination:

$$R^2 = 1 - \frac{\sum_{i=1}^{n}(y_i - \hat{y}_i)^2}{\sum_{i=1}^{n}(y_i - \overline{y})^2} = 1 - \frac{s_{\hat{y}}^2}{s_{\hat{y}}^2} = 1 - 1 = 0$$

Further the coefficient of determination R^2 is the squared value of the multiple correlation coefficient R. Therefore the coefficient of determination is a measure for the linear relationship, too. Values of R^2 in the interval $(0; 0.25)$ indicate a weak correlation. Values of R^2 in the interval $(0.25; 0.64)$ indicate a moderate correlation. Values of R^2 in the interval $(0.64; 1)$ indicate a strong correlation. R^2 is as well as R.

8.1.4 Adjusted R-Square

How can we tell how many independent variables should be added to the model? The intention is to select only a few independent variables X_1, X_2, ..., X_p to explain the value of Y.

Problem: The values of R and R^2 respectively, will never decrease if a further independent variable is added to the linear model $Y = b_0 + b_1 \cdot x_1 + b_2 \cdot x_2 + ... + b_p \cdot x_p$, even if the further independent variable has no influence on the value of Y. The measure **adjusted R-Square** $R_a^2 \in (-\infty; 1]$ indicates whether an additional independent is helpful in the linear model. The measure R_a^2 is defined as:

$$R_a^2 = 1 - (1 - R^2)\frac{n-1}{n-p-1} \text{ for } p \leq n-2$$

with n is the sample size and p is the number of independent variables. The value of R_a^2 may decrease if an additional independent variable has no influence on the value of Y, but the value of R^2 would be unchanged.

Example 8.12
Let us consider the model $Y \approx b_0 + b_1 \cdot x_1 + b_2 \cdot x_2$ with the following data set ($n = 5$):

y	x_1	x_2
1	10	−8
2	−12	9
3	13	12
5	−15	−16
8	18	16

Wet get the following values:

$$R = 0.307 \text{ multiple correlation coefficient;}$$

this means weak correlation

$$R^2 = 0.094 \; R\text{-Square/measure of determination}$$
$$R_{\text{a}}^2 = -0.812 \text{ adjusted } R\text{-Square}$$

The value of R_{a}^2 is inadequate.

Adjusted R^2 should only be used as an indicator of whether a new independent variable should be added in the model or not. If the value of R_{a}^2 decreases while adding another independent variable to the model, this is a hint, that the model with this additional independent variable may not be better than the model without this variable.

Example 8.13 (*Miles_Per_Gallon.sav* c.f. Berenson et al. [2015] p. 671)
Is it a good or bad idea to add the independent variable "Horsepower" to the model $\boxed{\text{MPG}} \approx b_0 + b_1 \cdot \boxed{\text{Weight}}$?
The correlation between MPG and weight is $r = 0.825$; therefore we get the following value of the adjusted R-Square R_a^2:

$$R_{\text{a}}^2 = 1 - (1 - R^2)\frac{n-1}{n-p-1} = 1 - (1 - 0.825^2) \cdot \frac{50-1}{50-1-1}$$
$$= 1 - 0.319 \cdot \frac{49}{48} = 0.674$$

If we add the independent variable horsepower to the model $\text{MPG} \approx b_0 + b_1 \cdot$ weight $+ b_2 \cdot$ horsepower, we get the following value of R_a^2:

$$R_a^2 = 1 - (1 - R^2)\frac{n-1}{n-p-1} = 1 - (1 - 0.866^2) \cdot \frac{50-1}{50-2-1}$$
$$= 1 - 0.250 \cdot \frac{49}{47} = 0.739$$

Thus the value of R_a^2 had changed from 0.674 to 0.739; this means that the change of R_a^2 indicates that the horsepower has an impact on the gasoline mileage.

Result: Take R or R^2 as a measure of the strength of correlation. Take R_a^2 as an indicator whether an additional independent variable of the model will improve the model or not.
The values of R^2 are all lying in the interval [0;1], but the values of R_a^2 may be (absurdly) negative.

8.1.5 Multicollinearity

An essential assumption of every linear model is that the independent variables $X_1, X_2, \ldots X_p$ do not depend linearly on one another, i.e. that they are not **collinear**. The linear independence can be seen from the multiple correlation coefficient (strong correlation). Indeed, if an independent variable can be represented as a linear combination of the other independent variables, even slight roundings can lead to considerable errors for the predicted values.
If two or more of the exogenous variables X_i are not stochastically independent, we say that we have **collinearity** or **multicollinearity**. A measure of collinearity is the so called **variance inflation factor** VIF:

$$VIF = \frac{1}{1 - R_i^2}$$

with R_i is the value of the multiple correlation coefficient for the model that X_i is a linear combination of the rest of the independent variables:

$$X_i = b_0 + \sum_{i \neq j} b_j X_j$$

In case of a strong correlation like $R_i \geq 0.95$, we get:

$$R_i \quad \geq \quad 0.95$$

$$R_i^2 \quad \geq \quad 0.9025 \approx 0.9$$

$$1 - R_i^2 \quad \leq \quad 0.1$$

$$\frac{1}{1 - R_i^2} \quad \geq \quad \frac{1}{0.1} = 10$$

This means that values of VIF larger or equal ten indicate a present collinearity. In case of a perfect collinearity, i.e. $R_i = 1$, the value of VIF is not defined.

Example 8.14
If we want to explain the rent Y of a flat by the square meters X_1 of the flat and the number of rooms X_2 of the flat. The two independent variables X_1, X_2 depend on each other. The variance inflation factor is larger than 10. So we have to deselect one of these variables X_1 or X_2 from the model.

As an index of collinearity SPSS computes the so called **tolerance** $1 - R_i^2$, too. Values of the tolerance less or equal to 0.1 indicate a present collinearity.

Result: In case of collinearity ($VIF = \dfrac{1}{1 - R_i^2} \geq 10$) the independent variable X_i should be deleted in the model.

Example 8.15 (*Miles_Per_Gallon.sav* see Berenson et al. [2015] p. 671)
The correlation of the model "Horsepower $\approx b_0 + b_1 \cdot$ Weight" is $r = 0.743$. So the *VIF*-values of the weight and of the horsepower is $1/(1 - 0.743^2) \approx 2.23$; so the value of *VIF* is not larger than ten; this means that there is no collinearity. Furthermore the tolerances of the weight and of the horsepower are $1 - 0.743^2 \approx 0.45$ and not smaller than 0.1.

8.1.6 Forecasting

We want to forecast a value based on a multiple linear regression model.

Example 8.16 (*Miles_Per_Gallon.sav* c.f. Berenson et al. [2015] p. 671)
We assume a linear relationship between MPG, Weight and Horsepower:

$$\boxed{\text{MPG}} \approx b_0 + b_1 \cdot \boxed{\text{Weight}} + b_2 \cdot \boxed{\text{Horsepower}}$$

resp.:

$$\text{Model: } Y \approx b_0 + b_1 \cdot x_1 + b_2 \cdot x_2$$

The least squares method estimates the constant and the regression coefficients as follows:

$$\widehat{b_0} = 58.157$$
$$\widehat{b_1} = -0.007$$
$$\widehat{b_2} = -0.118$$

Interpretation of the regression coefficient b_1: the value -0.007 is an estimated value of the expected decrease in gasoline mileage corresponding to an increase of one pound in the weight when the horsepower is unchanged.

Interpretation of the regression coefficient b_2: the value -0.118 is an estimated value of the expected decrease in gasoline mileage corresponding to an increase of one unit in the horsepower when the weight is unchanged.

We predict values of Y based on the regression model.

With 2 500 pounds of weight and with 80 unit as horsepower we get the predicted gasoline mileage of the car as follows:

$$58.157 - 0.007 \cdot 2\,500 - 0.118 \cdot 80 = 31.6$$

This means that the predicted miles per gallon are about 31.6 miles.
The value 31.6 miles is an interpolated value because the value 2 500 is lying

in the interval [1 755; 4 360] and the value 80 is lying in the interval [48; 165].

To check whether the interpolated value 31.6 miles is a good or bad predicted value we calculate the correlation between gasoline mileage, weight and horsepower. The multiple correlation between gasoline mileage, weight and horsepower is 0.866 (see example 8.11), i.e. strong correlation. Therefore the predicted interpolated value of 31.6 miles is reliable due to the strong correlation.

8.1.7   Heteroscedasticity

We have to check homoscedasticity in the model $Y \approx b_0 + b_1 \cdot x_1 + \ldots + b_p \cdot x_p$. In case of two or more independent variables it is not possible to construct a two dimensional scatter diagram between the dependent variable and all independent variables.

However, in case of heteroscedasticity the residuals $y - (\widehat{b}_0 + \widehat{b}_1 \cdot x_1 + \widehat{b}_p \cdot x_p)$ increase. Therefore we check the required homoscedasticity with a two dimensional scatter plot between the standardized predicted values (zpred) and the standardized residulas (zresid). This scatter plot is called **residuum plot**.

For homoscedasticity the points in the diagram of the residuals should look as if they were plotted at random. If e.g. there is a v-formation, we have heteroscedasticity.

Example 8.17 (*Miles_Per_Gallon.sav* see Berenson et al. [2015] p. 671)
In the model $\boxed{Y \approx b_0 + b_1 \cdot x_1 + b_2 \cdot x_2}$ the scatterplot between the standardized residuals "zresid" and the standardized predicted values "zpred" looks like:

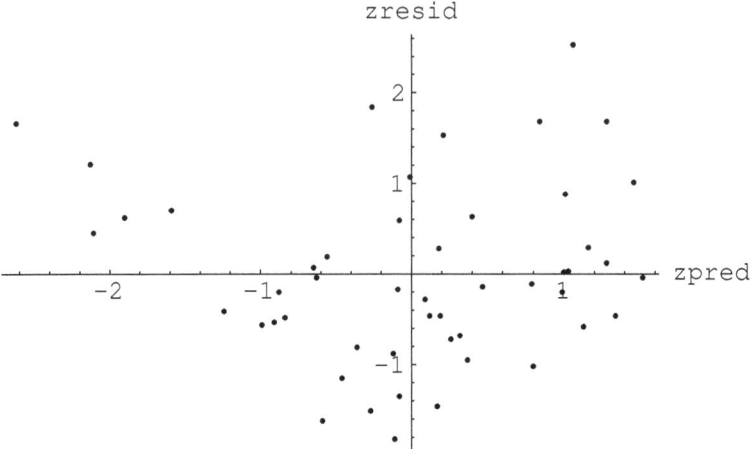

Residuum plot

In the scatterplot there is no v-formation; this means that the values of b_0, b_1, b_2 may be estimated due to the least squares method, because we have homoscedasticity.

8.1.8   Testing in Linear Regression Models

Preliminary remark on the terms "stochastically independent" and "uncorrelated": Two discrete random variables X, Y are called **stochastically independent**, if: $P(X = x \cap Y = y) = P(X = x) \cdot P(Y = y)$ for all x, y. Two random variables X, Y are called **uncorrelated**, if: $Cov[X, Y] = 0$. The theoretical covariance will be calculated as follows: $Cov[X, Y] = E[X \cdot Y] - E[X] \cdot E[Y]$.

The two terms are summarized as follows: If two random variables are stochastically independent, so they are also uncorrelated. Uncorrelation is a weaker one property as stochastic independence.

If two random variables are uncorrelated, it cannot generally (exception: Normally distributed random variables) concluded that the two random variables are stochastically independent.

116

Example 8.18

Let X, Y be two discrete random variables with the following probability function:

	X				
Y	-2	-1	1	2	Σ
-2	1/10	0	0	0	1/10
-1	0	0	4/10	0	4/10
1	0	4/10	0	0	4/10
2	0	0	0	1/10	1/10
Σ	1/10	4/10	4/10	1/10	1

The expected means are:

$$E[X] = (-2) \cdot \frac{1}{10} + (-1) \cdot \frac{4}{10} + 1 \cdot \frac{4}{10} + 2 \cdot \frac{1}{10} = 0$$

$$E[Y] = (-2) \cdot \frac{1}{10} + (-1) \cdot \frac{4}{10} + 1 \cdot \frac{4}{10} + 2 \cdot \frac{1}{10} = 0$$

$$E[X \cdot Y] = (-2) \cdot (2) \cdot \frac{1}{10} + (-1) \cdot 1 \cdot \frac{4}{10} + 1 \cdot (1-) \cdot \frac{4}{10} + 2 \cdot 2 \cdot \frac{1}{10} = 0$$

X, Y are uncorrelated

$$Cov[X, Y] = 0 - 0 \cdot 0 = 0$$

But X, Y are not stochastically independent, for example:

$$P(X = 1 \cap Y = -1) = \frac{4}{10} \neq \frac{4}{10} \cdot \frac{4}{10} = P(X = 1) \cdot P(Y = -1)$$

To run a test in a multiple egression model we must assume that the residuals ε_i = "observed value minus predicted value" are uncorrelated and have Normal distribution:

$$\varepsilon_i \sim N(0; \sigma)$$
$$Cov(\varepsilon_i \cdot \varepsilon_j) = 0 \text{ for } i \neq j \, (\text{pairwise uncorrelated})$$

Due to the Central Limit Theorem the assumption of the Normal distribution of the residuals is negligible if the sample size is equal to or larger than 30, i..e. $n \geq 30$ (c.f. Schlittgen [2009], Chapter 7, p. 157). For small sample sizes we check the Normal distribution of the residuals with a goodness-of-fit test, for large sample sizes we check the Normal distribution of the residulas with a QQ-Plot.

In particular, it is required that each of the n residual variable ϵ_i is Normally distributed. Since there is only one observation value for every residual variable, that is, the actual residuals, we check whether the n observed residuals are taken as a sample from a Normal distribution.

Example 8.19 (*Miles_Per_Gallon.sav* see Berenson et al. [2015] p. 671) The residuals in the model:

$$\boxed{\text{MPG} \approx b_0 + b_1 \cdot \text{weight} + b_2 \cdot \text{horsepower}}$$

have Normal distribution because the Histogram is fitted to the bell curve. Furthermore the loops in the Plot-Point-Diagram and in the Quantil-Quantil-Plot are lying close to the 45-degree-line.
The p-value of the Lilliefors test is at least 0.2 and the p-value of the Shapiro-Wilk test is exactly 0.372, so both tests confirm the Normal distribution of the residuals.

There is no option to check with SPSS whether the residuals are uncorrelated or not. The Durbin-Watson test checks autocorrelation with the lag one for a time series model, that is the Durbin-Watson test checks the correlation between ε_t and ε_{t-1}, t=time.

Which predictors contribute to the regression model? We want to check, which independent variable contributes much to the model. Perhaps there are too many predictors in the model. To check the influence of an independent variable we consider the following testing problem:

$$H_0 : b_i = 0 \text{ versus } H_1 : b_i \neq 0$$

The test statistic of this testing problem is t-distributed. Therefore this test is called t-test.

Example 8.20 (*Miles_Per_Gallon.sav* see Berenson et al. [2015] p. 671)
In the model $Y = b_0 + b_1 x_1 + b_2 x_2 + \varepsilon$ with the variables:

$$
\begin{aligned}
Y &= \text{gasoline mileage} \\
x_1 &= \text{weight} \\
x_2 &= \text{horsepower}
\end{aligned}
$$

we assume Normal distribution of the residuals ε = observed miles minus predicted miles per gallon.

We want to check whether the independent variable x_1 =weight is necessary in the model or not:

---

t-test in a regression model

H_0: The weight has no influence on the miles per gallon
   (i.e. $b_1 = 0$)

versus

H_1: The weight has an influence on the miles per gallon
   (i.e. $b_1 \neq 0$)

Rejection of $H_0 \Leftrightarrow p$-value $\leq \alpha$

⚠ If the test rejects the null hypothesis H_0 the influence of the weight is called **significant** influence.

SPSS computes the p-value as 0.00001. So the p-value is smaller than 0.05; this means that we reject the null hypothesis; this means the weight has a significant influence on the miles per gallon.

Does the horsepower impacts the miles per gallon?

We consider the following testing-problem:

t-test in a regression model

H_0:   The horsepower has no influence on the miles per gallon (i.e. $b_2 = 0$)

versus

H_1:   The horsepower has an influence on the miles per gallon (i.e. $b_1 \neq 0$)

Rejection of H_0 if and only if: p-value ≤ 0.05

SPSS computes the p-value as 0.001. So the p-value is smaller than 0.05; this means that we reject the null hypothesis; this means that the horsepower has an significant influence on the miles per gallon; this means that the variable horsepower contributes much to the model.

There are two small p-values, indicating that these variables contribute much to the model.

We want to check whether we need the constant b_0 in the model $Y = b_0 + b_1 x_1 + b_2 x_2 + \varepsilon$:

t-test in a regression model

H_0 :   The constant equals zero (i.e. $b_0 = 0$)

versus

H_1 :   The constant unequals zero (i.e. $b_0 \neq 0$)

Rejection of H_0 if and only if the p-value ≤ 0.05

SPSS computes the p-value as $3.6 \cdot 10^{-26} \approx 0$. Therefore the p-value is smaller than 0.05; so we reject the null hypothesis; this means that the constant b_0 is necessary in the model.

We want to check the acceptability of the regression model, this means we test whether we need all independent variables $x_1, x_2, \ldots, x_p$ simultaneously:

$$H_0 : b_1 = b_2 = \ldots = b_p = 0$$
versus
$$H_1 : \text{"At least one } b_i \text{ is unequal zero"}$$

The test statistic is F-distributed. Therefore this test is called F-test.

Example 8.21 (*Miles_Per_Gallon.sav* c.f. Berenson et al. [2015] p. 671)
Is the model $\boxed{Y \approx b_0 + b_1 \cdot x_1 + b_2 \cdot x_2}$ acceptable?

We consider the following testing problem:

F-test in a regression model
H_0:   The model is not acceptable (i.e. $b_1 = b_2 = 0$)
versus
H_1:   The model is acceptable (i.e. at least one
         b_i is unequal zero)
Rejection of H_0 if and only if: p-value ≤ 0.05

SPSS computes the p-value as $8 \cdot 10^{-15} \approx 0$ and the p-value is listed in the ANOVA-Table. So the p-value is smaller than $\alpha = 0.05$, so we reject the null hypothesis; this means that the gasoline mileage depends on a linear combination of some predictors. The gasoline mileage depends not on some chance variations.

8.1.9   Summary

The purpose of a multiple linear regression model $Y \approx b_0 + b_1 \cdot x_1 + b_2 \cdot x_2 + b_3 \cdot x_3 + \dots b_p \cdot x_p$ is to predict values of Y.

The measures of an adequate multiple linear regression model should have the following values:

Measure	Range
R	> 0.8
VIF	< 10
Tolerance	> 0.1

Furthermore, the points in the residual plot should be displayed at random (no systematic pattern) i.e. homoscedasticity, to provide the validity of the least squares method.

For the linear regression model $Y \approx b_0 + b_1 \cdot x_1 + \ldots + b_p \cdot x_p$ there are two kinds of tests.

The p-value in the ANOVA-table indicates whether the considered regression model is adequate or not.

The p-values in the Coefficients-table indicate which independent variable has a significant influence on the dependent variable.

We have considered the following statistical inferences:

1) The estimated values $\widehat{b}_0, \widehat{b}_1, \widehat{b}_2, \ldots \widehat{b}_p$ of the regression coefficients are computed by the least squares method.

2) If the multiple correlation coefficient $R \in [0; 1]$ indicates weak or medium correlation, the model is bad.

3) The correlation coefficient of Bravais-Pearson $r \in [-1; 1]$ shows the correlation of two variables only.

4) If there is a v-formation in the residuum plot, i.e. heteroscedasticity, the estimated values of $\widehat{b}_0, \widehat{b}_1, \widehat{b}_2, \ldots \widehat{b}_p$ are incorrect.
   In case of homoscedasticity the points in the residuum plot are looking like plotted by random.

5) For VIF ≥ 10 we have multicollinearity this means that the estimated values of $\widehat{b}_0, \widehat{b}_1, \widehat{b}_2, \ldots \widehat{b}_p$ are incorrect.

6) A predicted value y is an extrapolated value if $x \notin [x_{min}; x_{max}]$.

7) A predicted value y is an interpolated value if $x \in [x_{min}; x_{max}]$.

8) Extrapolated predicted values are never reliable. Interpolated predicted values are reliable, if the multiple correlation coefficient indicates a strong correlation.

9) The leverage value indicates whether the observed value has a low or a high impact to the shape of the regression line. An outlier with a high leverage value should be removed.

10) If we add another independent variable to the model and the value of R_a^2 is increasing this indicates that the additional independent variable should be added to the model.

11) The residuals are the difference observed value minus predicted value.

12) If in the chart "Quantile-Quantile Diagram" the points are plotted close to the 45-degree line we conclude that the residuals have Normal distribution.

13) p-value of the t-test $\leq 0.05 \Leftrightarrow X_i$ has a significant influence on Y.

14) p-value ANOVA $\leq 0.05 \Leftrightarrow$ the model has sense.

8.1.10 SPSS-Commands Multiple Linear Regression

We consider the example 8.1.

Scatterplot Without Regression Line

1. Option:

1) Graphs → Legacy Dialogs → Scatter/Dot ......

2) Select "Simple Scatter".
   Click "Define".

3) Y-Axis = MPG
   X-Axis = Weight

4) Click "ok".

2. Option:

1) Graphs → Chart Builder → ok

2) Choose from gallery: Scatter/Dot

3) Double click on the diagram "Simple Scatter"

4) *x*-Axis = weight
   y-Axis = MPG

5) (If you want a point-identification in the scatter diagram you have to select "Groups/Point ID", select "Point ID label", Point Label Variable = (name of the variable))

6) ok

A scatterplot that was constructed with the second option can be processed further.

Scatterplot With Regression Line

1) Analyze → Regression → Curve Estimation

2) Dependent(s) = MPG
   Variable = weight
   Models = Linear

3) Click "ok".

Partial Correlation

1) Analyze → Correlate → Partial …

2) Variables = weight and horsepower
   Controlling for = MPG

3) Click "ok".

The partial correlation coefficient between weight and horsepower is 0.264.

Linear Regression

1) In the SPSS Data-View open the file "Miles_per_Gallon.sav".

2) Analyze → Regression → Linear ...

3) Dependent= "MPG"
   Independent(s) = "weight" and "horsepower"
   Method=Enter

4) Click "Statistics".
   Select "Estimates" in "Regression Coefficient".
   Select "Model fit" (=Default).
   Select "Part and partial correlations" and "Collinearity diagnostics".
   Click "Continue".

5) Click "Plots".
   Select as variables:
           Y: ∗ZRESID
           X: ∗ZPRED
   Select "Histogram" and "Normal probability plot". Click "Continue".

6) Click "Save".
   Select "Unstandardized" in the Predicted Values group.
   Select "Unstandardized" in the Residuals group.
   Select "Cook's" and "Leverage values" in the Distances group.
   Click "Continue".

7) Click "OK".

The table "Model Summary" shows the multiple correlation coefficient $R = 0.866$.

Model Summary^b

Model	R	R-Square	Adjusted R-Square	Std Error of the Estimate
1	.866^a	.749	.739	4.1766

a. Predictors : (Constant), Horsepower, weight
b. Dependent Variable: MPG

The table "ANOVA" shows in the column (empirical) Significance the p-value=0.000 of the testing problem $H_0 : b_1 = b_2 = \ldots = b_p$.

ANOVA[b]

Model		Sum of Squares	df	Mean Square	F	Sig.
1	Regression	2 451.974	2	1 225.987	70.281	.000[a]
	Residual	819.868	47	17.444		
	Total	3 271.842	49			

a. Predictors: (Constant), Horsepower, Weight
b. Dependent Variable: MPG

The correlations -0.825 resp. -0.788 between the dependent variable (here: MPG) and the independent variable (here: either weight or horsepower) are listed in the table "Coefficients" in the column "Correlations" in the column "Zero-order".

The p-value 0 of the testing problem $H_0 : b_0 = 0$, the p-value 0.00001 of the testing problem $H_0 : b_1 = 0$ and the p-value 0.001 of the testing problem $H_0 : b_2 = 0$ are listed in the SPSS-Table "Coefficients" in the column (empirical) significance.

The p-value of 0.00001 shows us that we reject the null hypothesis $H_0 : b_1 = 0$; this means that the weight is a significant predictor for MPG.

The p-value of 0.001 shows us that we reject the null hypothesis $H_0 : b_2 = 0$; this means that the horsepower is a significant predictor for MPG.

In the column "Unstandardized Coefficients" column B of the table "Coefficients" the estimators 58.157, -0.007, -0.118 for the constant and the regression coefficients are shown.

The values of the residuals $y_i - (\widehat{b}_0 + \widehat{b}_1 x_{i1} + \ldots + \widehat{b}_p x_{ip})$ are shown in the Data-View in the column RES_1. The residuals are the differences between the observed values and the model-predicted values of the dependent variable.

The predicted values of the model are listed in the SPSS-Data-View in the column PRE_1.

Forecasting

Enter 2 500 weight as the fifty-first value and 80 horsepower as the fifty-first value in the Data-View. After this, please let SPSS run again a linear regression. Then the predicted value of 31.6 miles of this car is listed in the fifty-first row in the column PRE_1 in the Data-View.

Output

Coefficients[a]

Model		Unstandardized Coefficients		Standardized Coefficients	t	Sig.	Correlations			Collinearity statistic	
		B	Std Error	Beta			Zero-order	Partial	Part	Tolerance	VIF
1	(Constant)	58.157	2.658		21.878	.000					
	Weight	−.007	.001	−.534	−4.903	.000	−.825	−.582	−.358	.450	2.224
	Horsepower	−.118	.033	−.392	−3.600	.001	−.788	−.465	−.263	.450	2.224

a. Dependent variable: MPG

128

8.2 Binary Logistic Regression

The dependent variable Y must be leveled as follows:

nominal:	yes, but only binary variables
ordinal:	no
scale:	no

The independent variables $X_1, \ldots, X_p$ must be leveled as follows:

nominal:	yes
ordinal:	yes
scale:	yes

We want to forecast the value $Y = 1$ or $Y = 0$ of the binary variable Y based on the observed values $x_1, x_2, \ldots, x_p$.

Example 8.22 (*Kredit_biLogReg.sav* c.f. Handl page 10)
The total costs X_1 (in monetary units) were observed of 20 local branches of a bank in Baden-Württemberg. The 20 local branches are discriminated into two types: Type I are local branches with a high market share and type II are local branches with a low market share. The types of a branch are coded as 1 for type I and 0 for type II:

No.	Total costs	Type	No.	Total costs	Type
1	478.2	1	11	413.8	1
2	247.3	1	12	379.7	1
3	223.6	1	13	400.5	1
4	505.6	1	14	404.1	1
5	399.3	1	15	499.4	0
6	276.0	1	16	674.9	0
7	542.5	1	17	468.6	0
8	308.9	1	18	601.5	0
9	453.6	1	19	578.8	0
10	430.2	1	20	641.5	0

We want to forecast the type of a new local branch based on the value of the total costs $x_1 = 520$. The binary logistic regression estimates for every branch the probability of type I, i.e. $P(Y = 1)$. An estimated probability larger than 0.5 indicates type I. And an estimated probability not larger than 0.5 indicates type II.

A simple approach is to estimate the probability $P(Y = 1 \mid x_1 = 520)$ due to a linear combination z of the independent variable(s):

$$z = \text{Constant} + b_1 \cdot \text{total costs}$$
$$z = b_0 + b_1 \cdot x_1$$

Problem: Such probabilities $z = \widehat{P}(Y = 1 \mid x_1)$ may be outside of the interval $[0;1]$. If we take the logistic distribution instead of the linear combination we get probabilities lying in the interval $[0;1]$:

$$\widehat{P}(Y = 1 \mid x_1) = \frac{e^z}{1 + e^z} = \frac{1}{1 + e^{-z}}$$

The predicted types (listed in SPSS in the column PGR=predicted group) of the 20 local branches due to the binary logistic regression are:

No.	Total costs	Type	Pred. type	No.	Total costs	Type	Pred. type
1	478.2	1	1	11	413.8	1	1
2	247.3	1	1	12	379.7	1	1
3	223.6	1	1	13	400.5	1	1
4	505.6	1	1	14	404.1	1	1
5	399.3	1	1	15	499.4	0	1
6	276.0	1	1	16	674.9	0	0
7	542.5	1	0	17	468.6	0	1
8	308.9	1	1	18	601.5	0	0
9	453.6	1	1	19	578.8	0	0
10	430.2	1	1	20	641.5	0	0

Sometimes the observed type and the predicted type are different. Fourteen times the observed type is type I, and for thirteen times the observed type I is the same as the predicted type, this means that $1 - \frac{1}{14} = 92.9\%$ of the predicted types are correct. And six times the observed type is type II and for two times the predicted type differs from the type II, this means that $1 - \frac{2}{6} = 66.7\%$ of the predicted types are correct. Overall 17 predicted types of 20 types are correct, i.e. 85%.

The Hosmer-Lemeshow test verifies the goodness-of-fit of the model $Y \approx$

$b_0 + b_1 \cdot x_1$:

> **Hosmer-Lemeshow-Goodness-of-fit test**
> H_0: The model has a good fit
> versus
> H_1: The model has no good fit
> Rejection of $H_0 \Leftrightarrow p$-value ≤ 0.05

The p-value of the Hosmer-Lemeshow-goodness-of-fit test in the last step is 0.762; no rejection of H_0; the model has a good fit.

The Wald test is testing the effect of an independent variable onto the dependent variable:

> **Wald test**
> $H_0 : b_i = 0$ (the i-th variable has no effect)
> versus
> $H_1 : b_i \neq 0$ (the i-th variable has an effect)
> Rejection of $H_0 \Leftrightarrow p$-value ≤ 0.05

For the regression coefficient b_1 we get the p-value of the Wald tests as 0.035; H_0 is rejected; this means that X_1 has a significant effect onto Y; i.e. the total costs have a significant effect on the type of the branch.

The estimated regression coefficients of the model $b_0 + b_1 \cdot x_1$ are $\widehat{b}_0 = 15.219$ and $\widehat{b}_1 = -0.029$. The predicted type of a new local branch with 520 total costs is type II; this means that the predicted market share of the new local branch is low.

8.2.1 SPSS-Commands Binary Logistic Regression

1) Please open the file "'Kredit_biLogReg.sav'".

2) Analyze → Regression → Binary Logistic ...

3) Dependent = "Typ"
   Covariates = "Gesamtkosten"

Method = Enter

4) If there are independent dichotomous or nominal leveled variables click "Categorical" and select all independent nominal leveled variables as Categorical Covariates. Select "Continue".

5) Click "Save ... ".
Under "Predicted Values"' select "'Group membership".
(Select "Probabilities" to get the estimated probabilities $P(Y = 1)$)
Select "Continue".

6) Click "Options" ... ".
Under "Statistics and Plots" select "Hosmer-Lemeshow-goodness-of-fit".
Select "Continue".

7) Select "OK".

The value of Nagelkerke R-Square is 0.713; that indicates a strong correlation of the model:

Model Summary

Step	-2 Log likelihood	Cox- & Snell R Square	Nagelkerke R Square
1	10.457a	0.503	0.713

a. Estimation terminated at iteration number 7 because parameter estimates changed by less than 0.001.

The p-value 0.762 of the Hosmer-Lemeshow-goodness-of-fit test indicates a good fit of the model:

Hosmer and Lemeshow test

Step	Chi-square	df	Sig.
1	4.959	8	0.762

The values of the regression coefficients are listed in the table in the column "B" of the table "Variables in the Equation".

Variables in the Equation

		B	S.E.	Wald	df	Sig.	Exp(B)
Step 1[a]	Gesamtkosten	-.029	.014	4.422	1	.035	.971
	Constant	15.219	7.022	4.698	1	.030	4069247.7

a. Variable(s) entered on step 1: Gesamtkosten.

The comment of the sign of the estimated regression coefficient $\hat{b}_1 = -0.029$ is as follows: The higher the total costs, the more likely the branch will have a predicted type 0, i.e. the more likely a branch with high total costs will have a low market share.

The predicted types of the binary logistic regression are listed in the column PGR_1 (abbreviation for predicted group) of the Data-View.

Over all 85.0% of the observed types are predicted properly:

133

Classification Table[a]

		Predicted		
		Type		Percentage
Observed		0	1	Correct
Step 1	Type 0	4	2	66.7
	1	1	13	92.9
Overall Percentage				85.0

a. The cut value is 0.500

Forecasting

Enter the value 520 in the Data-View in the column "Costs". Let SPSS run again a binary logistic regression. The predicted type 0, i.e. type II, of the new local branch with total costs of 520 is listed in the column PGR_2 in the Data-View.

8.3 Multinomial Logistic Regression

The dependent variable Y must be leveled as follows:

nominal:	yes
ordinal:	no
scale:	no

The independent variables $X_1, \ldots, X_p$ must be leveled as follows:

nominal:	yes
ordinal:	yes
scale:	no

We want to forecast the value of the nominal leveled variable Y based on the observed values $x_1, x_2, \ldots, x_p$.

Example 8.23 (*cereal.sav* c.f. SPSS-Tutorial of IBM)
For selective advertising of breakfast products a corporation has asked 880 persons about:

$Y=$ preferred breakfast
1=breakfast bar, 2=oat meal, 3=cereal

$X_1=$ age category
1=under 31, 2=31 - 45, 3=46 - 60, 4=over 60

$X_2=$ active lifestyle (no=0, yes=1)
yes=at least two times sports every week

$X_3=$ gender (0=male, 1=female)

The multinomial logistic regression analysis estimates the probabilities of the preferred breakfast category k with $k = 1,2,3$ with the logistic distribution:

$$\frac{e^{z_k}}{e^{z_1} + e^{z_2} + e^{z_3}} \; ; k = 1,2,3$$

here $z_k = b_{k0} + b_{k1} \cdot x_1 + b_{k2} \cdot x_2 \; ; k = 1,2,3$ is a linear combination of the independent variables X_1, X_2.

For every respondent we get three probabilities of the three breakfast categories bar, oat, cereal. The predicted breakfast is the breakfast with the largest probability. The predicted breakfast of a respondent is basing upon his age and his lifestyle.

Two different goodness-of-fit tests checks the goodness of the model $Y \approx b_0 + b_1 \cdot x_1 + b_2 \cdot x_2$:

Chi-Square Goodness-of-Fit test of Pearson
H_0: Good fit of the model
versus
H_1: No good fit of the model
Rejection of $H_0 \Leftrightarrow p$-value ≤ 0.05

The p-value of the Chi-Square Goodness-of-Fit test of Pearson in the last step is 0.641; no rejection of H_0; good fit of the model. The p-value of the Chi-Square Goodness-of-Fit test in the last step is 0.472; no rejection of H_0; good fit of the model.

The Likelihood Ratio Test verifies the effect of the independent variables onto the dependent variable:

The p-value of the Likelihood Ratio Test for the variable "age category" is about 0; this means rejection of H_0; this means the independent variable "age category" has an effect onto the dependent variable preferred breakfast.

The p-value of the Likelihood Ratio Test for the variable "active lifestyle" is about 0; this means rejection of H_0; this means the independent variable "active lifestyle" has an effect onto the dependent variable preferred breakfast.

The Wald test verifies what category of an independent variable has an effect onto the categories of the dependent variable:

Wald Test
H_0: The independent variable has no effect onto the dependent variable
versus
H_1: The independent variable has an effect onto the dependent variable
Rejection of $H_0 \Leftrightarrow p$-value ≤ 0.05

The p-values of the Wald tests for the three categories of the dependent variable Y are:

	Category		
	Breakfast bar	Oat meal	Cereal
	p-value	p-value	p-value
Constant	0.010	0.000	
Age category 1	0.003	0.000	
Age category 2	0.001	0.000	
Age category 3	0.428	0.000	
Age category 4	.	.	
No active lifestyle	0.000	0.342	
Active lifestyle	.	.	

The p-value of the Wald test of the category "breakfast bar" in the first age category is 0.003; i.e. rejection of H_0; this means the first age category has a significant effect onto the preferred breakfast category "breakfast bar". The p-value of the Wald test of the category "breakfast bar" in the third age category is 0.428; i.e. no rejection of H_0; this means that the third age category has no significant effect onto the preferred breakfast category "breakfast bar". The p-value of the Wald test of the category "Oat meal" in the inactive lifestyle category is 0.342; i.e. no rejection of H_0; this means that the inactive lifestyle category has no significant effect onto the preferred breakfast category "oat meal".

The regression coefficients are:

	Breakfast bar	Oat meal	Cereal
Category	B	B	B
Constant	−0.744	1.022	
Age category 1	0.938	−4.256	
Age category 2	1.047	−2.461	
Age category 3	0.263	−1.115	
Age category 4	0^b	0^b	
No active lifestyle	−0.786	0.178	
Active lifestyle	0^b	0^b	

b This parameter is set to zero because it is redundant.

The procedure of the multinomial logistic regression provides a predicted category of the preferred breakfast of an new customer based on his age and his lifestyle.

8.3.1 SPSS-Commands Multinomial Logistic Regression

1) Please open the file "cereal.sav".

2) Analyze → Regression → Multinomial Logistic ...

3) Dependent = Breakfast
   Factor(s)=   Age category
                Lifestyle
                Gender

4) Click "Model ... ".
   Specify Model=Custom/Stepwise
   Select in the dropdown menu "Build Terms"' the term "Main effects".
      Forced Entry Terms=   Age category
                            Lifestyle
   Click "Continue".

5) Click "Statistics ... ".
   Under "Model" select "Cell probabilities", "Classification table"', "Good-ness-of-fit".
   Under "Parameters" select "Estimates" with a 95%-Confidence inter-

val.
Click "Continue".

6) Click "Save ... ".
   Under "Saved variables" select "Predicted category". (Select "Pre-
   dicted category probability".)
   Click "Continue".

7) Click "OK"

The predicted breakfast is listed in the column PRE_1 of the Data-View.
(The estimated probabilities of the predicted breakfast categories 1,2,3 are
listed in the column EST1_1, EST2_1, EST3_1 of the Data View.)

The p-values of the two Chi-Square-Goodness-of-Fit tests are 0.641 (Pearson) resp. 0.472 (Deviance); i.e. the model has a good fit:

Goodness-of-Fit

	Chi-Square	df	Sig.
Pearson	19.075	22	.641
Deviance	21.801	22	.472

The p-value of the Likelihood Ratio test are 0.000 (age) and 0.000 (lifestyle):

Likelihood Ratio Tests

Effect	Model Fitting Criteria -2 Log-Likelihood of Reduced Model	Likelihood Ratio Tests		
		Chi-Square	df	Sig.
Intercept	135.915[a]	.000	0	.
Altersklasse	451.066	315.151	6	.000
Lebensform	160.949	25.034	2	.000

The chi-square statistic is the difference in -2 log-likelihoods between the final model and a reduced model. The reduced model is formed by omitting an effect from the final model. The null hypothesis is that all parameters of that effect are 0.

a. This reduced model is equivalent to the final model because omitting the effect does not increase the degrees of freedom.

The total rate of correct predicted breakfast categories is (118+251+127)/880 = 56.4 %. That is poor. Especially 118 of the 231 observed preferred Breakfast Bar cases are correct predicted, i.e. 51.1 %. And 251 of the 310 observed preferred Oatmeal cases are correct predicted, i.e. 81.0 %. And 127 of the 339 observed preferred Cereal cases are correct predicted, i.e. 37.5 %.

Classification

Observed	Predicted			Percent Correct
	Breakfast Bar	Oatmeal	Cereal	
Breakfast Bar	118	34	79	51.1 %
Oatmeal	14	251	45	81.0 %
Cereal	96	116	127	37.5 %
Overall Percentage	25.9 %	45.6 %	28.5 %	56.4 %

Parameter Estimates

Preferred breakfast[a]		B	Std. Error	Wald	df	Sig.	Exp(B)	95% Confidence Interval for Exp(B)	
								Lower Bound	Upper Bound
Breakfast bar	Intercept	-.744	.287	6.707	1	.010			
	Alterskl.1	.938	.313	8.989	1	.003	2.555	1.384	4.719
	Alterskl. 2	1.047	.311	11.333	1	.001	2.848	1.549	5.239
	Alterskl. 3	.263	.332	.629	1	.428	1.301	.679	2.494
	Alterskl. 4	0[b]	.	.	0	.	.	.	.
	Lebensform nein	-.786	.181	18.945	1	.000	.456	.320	.649
	Lebensform ja	0[b]	.	.	0	.	.	.	.
Oat Meal	Intercept	1.022	.195	27.478	1	.000			
	Alterskl. 1	-4.256	.533	63.770	1	.000	.014	.005	.040
	Alterskl. 2	-2.461	.275	80.174	1	.000	.085	.050	.146
	Alterskl. 3	-1.115	.208	28.727	1	.000	.328	.218	.493
	Alterskl. 4	0[b]	.	.	0	.	.	.	.
	Lebensform nein	.178	.187	.902	1	.342	1.195	.828	1.724
	Lebensform ja	0[b]	.	.	0	.	.	.	.

a The reference category is cereal.
b This parameter is set to zero because it is redundant.

SPSS denotes the last category of the dependent variable and of the independent variable(s) as reference categories and these categories are not listed in the table of the Wald test.

Whether one of the first three categories of the independent variable $X_1 =$ "age category" has a significant effect onto the first category "breakfast bar" of the dependent variable $Y =$ "Breakfast" is shown in the SPSS-output "Parameter Estimates": The p-values 0.003 resp. 0.001 indicate a significant effect of the first two age categories onto the breakfast category; but the p-value of 0.428 of the third age category indicates no significant effect. Further the p-value 0.000 of the first category "not active" of the independent variable $X_2 =$ "lifestyle" indicates a significant effect onto the first category "breakfast bar" of the dependent variable $Y =$ "breakfast". And the p-values of about 0.000 resp. of the first three categories of the independent variable "age category" indicate a significant effect onto the second category "oat meal" of the dependent variable "breakfast". But the p-value 0.342 of the not active lifestyle category indicates no significant effect onto the breakfast category "oat meal".

Forecasting

We want to predict the preferred breakfast of a 48 years old woman with active lifestyle.
Enter the values agecat=3, gender=1 and active=1 in the 881st row in the Data-View. Let SPSS run one more time a multinomial logistic regression. The predicted preferred breakfast 3, i.e. cereal, is listed in the column PRE_2 in the Data-View

8.4   Ordinal Regression

The dependent variable Y must be leveled as follows:

nominal:	no
ordinal:	yes
scale:	no

The independent variables $X_1, \ldots, X_p$ must be leveled as follows:

nominal:	yes
ordinal:	yes
scale:	no

We want to forecast the value of the nominal leveled variable Y based on the observed values $x_1, x_2, \ldots, x_p$.

Example 8.24 (*german_credit.sav* c.f. SPSS-Tutorial of IBM)
A bank wants to predict the creditworthiness of a new customer based on the following values of the new customer:

$Y=$ Account-Status (chist), 1=no debt history, 2=no current debt, 3=payments current, 4=payments delayed, 5=critical account

$X_1=$ Number of existing credits (1,2,3,4) (numcred)

$X_2=$ other installment debts (othnstal)
1=bank, 2= stores, 3=none

$X_3=$ Housing (housng)
1=rent, 2=own, 3=free

$X_4=$ Age (in years) (age)

$X_5=$ Duration of a credit in month (duration)

The ordinal regression analysis estimates the probability of the predicted category of the ordinal leveled dependent variable. These probabilities are estimated based on the linear combination of the independent variables:

$$link(y_j) = \theta_j - [\beta_1 x_1 + \ldots + \beta_p x_p] \; ; j = 1, \ldots, J$$

where the link function is denoted by $link$. We have the choice between five link functions that we select according to the data. We consider the frequency distribution of Y Account status to select the matching link function:

Account Status

		Fre-quen-cy	Percent	Valid Per-cent	Cumu-lative Per-cent
Valid	No debt history	40	4.0	4.0	4.0
	No current debt	49	4.9	4.9	8.9
	Payments current	530	53.0	53.0	61.9
	Payments delayed	88	8.8	8.8	70.7
	Critical account	293	29.3	29.3	100.0
	Total	1000	100.0	100.0	

The last three categories 3,4,5 are more frequent than the first two categories. We have to select Complementary Log-Log as the link function.

There are several link functions for five special cases of data set of Y:

Function	Form	Typical application
Logit	$\log\left(\dfrac{\gamma}{1-\gamma}\right)$	Y has evenly distributed categories
Complementary log-log	$\log(-\log(1-\gamma))$	Higher categories of Y are more frequent
Negative log-log	$-\log(-\log(\gamma))$	Lower categories of Y are more frequent
Probit	$\Phi^{-1}(\gamma)$	Latent variable is Variable normally distributed
Cauchit (Inverse Cauchy)	$\tan(\pi(\gamma-0.5))$	Latent variable has many extreme values

What is it a latent variable? The latent variable is thought of as an unobserved continuous variable. When the value of this variable reaches a certain point, a "threshold", the observed response Y is "yes" for a special category of Y.

Because scale leveled variables are not allowed in the ordinal regression analysis we have to class the scale leveled independent variables X_1="Number of existing credits", X_4="Age" and X_5="duration in month" as follows:

X_1= Number of existing credits (numcred)
  1=none or only one existing credit, 2=mtwo or more existing credits

X_4= Age (in years) (age)
  1=up to 30 years, 2=31 up to 50 years, 3=over 50 years

X_5= Duration in month (duration)
  1=up to 24 months, 2= over 24 months

The recoded variables of X_1, X_4 and X_5 are now ordinal leveled. (SPSS offers the option of denoting scale leveled variables as covariates. But it is not possible to verify the goodness-of-fit of such a model, because many cells have frequencies of the value zero, so test is not applicable.)

Before running the ordinal regression procedure we have to reduce the number of categories of Y from five to three (if not we would get cells with frequencies zero and the goodness-of-fit test is not applicable):

Y= Account status
  1= no debt history and no current debts, 2=payments current, 3=payments delayed or critical account

The predicted category of the account status is based on the category of housing, other installment debts, age and the duration of existing payments and is listed in the column PRE_1 of the Data-View.

The Log-Likelihood-Chi-Square test verifies whether the model gives adequate predictions:

Log-Likelihood-Chi-Square Test
H_0 : The model gives no adequate predictions
H_1 : The model gives adequate predictions
Rejection of H_0 $\Leftrightarrow$ p-value ≤ 0.05

The p-value of the Log-Likelihood-Chi-Square test is about 0.000; this means that the model gives adequate predictions.

There are two test to check the goodness-of-fit of the model:

Chi-Square-Goodness-of-Fit test Pearson
H_0 : The observed data are consistent with the fitted model; good fit of the model
H_1 : The observed data are inconsistent with the fitted model; no good fit of the model
Rejection of $H_0 \Leftrightarrow p$-value ≤ 0.05

The p-value of the Chi-Square-Goodness-of-Fit test of Pearson is 0.000; we must conclude that we have a bad fit of the model.

Chi-Square-Goodness-of-Fit test of deviance
H_0 : The observed data are consistent with the fitted model; good fit of the model
H_1 : The observed data are inconsistent with the fitted model; no good fit of the model
Rejection of $H_0 \Leftrightarrow p$-value ≤ 0.05

The p-value of the Chi-Square-Goodness-of-Fit test due to the deviance is 0.000; we must conclude that we have a bad fit of the model.

The Wald test verifies for each category of the independent variables, whether these categories have a significant effect onto the dependent variable. SPSS denotes each last category of a independent or of the dependent variable as a reference category, so the last categories are redundant and missing in the table of the Wald test.

Wald test
H_0 : The i-th independent variable has no effect
versus
H_1 : The i-th independent variable has an effect
Rejection of $H_0 \Leftrightarrow p$-value ≤ 0.05

The category "none or only one exiting credit" has the p-value of about 0.000 and has a significant effect onto the account status. The category "other installment debts at the bank" has the p-value of about 0.000 and has a significant effect onto the account status. The first age class "under 31" has the p-value 0.028 and has a significant effect onto the account status. The category "other installment debts at stores" has the p-value 0.135 and has significant effect onto the account status. The housing categories "rent" and "own" have the p-values 0.387 resp. 0.644 and have no significant effect onto the account status. The second age category has no significant effect onto the account status due to its p-value of 0.883. The duration category "up to 24 months" has no significant effect onto the account status due to its p-value of 0.689.

8.4.1 SPSS-Commands Ordinal Regression

1) Please open the file "german_credit.sav".

2) Analyze → Regression → Ordinal ...

3) Dependent = chist_class
     Factor(s)=   numcred_class
                      other installment debts
                      Housing
                      age_class
                      duration_class

4) Click "'Options ... '".
    In the dropdown menu select as the "Link-Function" the function "Complementary log-log". Click "Continue".

5) Click "Output ... ".
    Under "Display" select "Goodness of fit statistics", "Summary statistics", "Parameter estimates" and "Test of parallel lines". Under "Saved Variables" select "Predicted category". (If you like you can select "Predicted category probability", too.) Under "Print Log-Likelihood" select "Including multinomial constant". Click "Continue".

6) Click "Location ...".
   Under "Specify Model" select "Main effects". Click "Continue".

7) Click "OK"

The first output is a warning about cells with zero frequencies:

Warnings

There are 77 (33%) cells (i.e., dependent variable levels by combinations of predictor variable values) with zero frequencies.

The upper bound is 20%, but with 33% there are too many zero frequencies. If we reduce the categories of Y from five to three we would get over 50% cells with zero frequencies.

We obtain 0.000 as the p-value of the Log-Likelihood-Chi-Square test, we conclude that the model gives adequate predictions:

Model Fitting Information

Model	−2 Log-Likelihood	Chi-Square	df	Sig.
Intercept only	866.054			
Final	495.375	370.679	8	.000

Link function: Complementary Log-log.

But the fit of the model is bad because we obtain the p-values 0.000 of the Chi-Square test of Pearson and based on the deviance:

Goodness-of-Fit

	Chi-Square	df	Sig.
Pearson	506.518	144	.000
Deviance	312.784	144	.000

Link function: Complementary Log-log.

Overall $(478+280)/1\,000 = 75.8\%$ of the predicted categories are the same as the observed categories of Y:

chist_class * Predicted Response Category Crosstabulation

Amount

Observed Category		Predicted Response Category		Total
		2	3	
chist_class	1	54	35	89
	2	478	52	530
	3	101	280	381
Total		633	367	1000

The estimators of the parameters are shown in the following table. Positive estimates mean a higher category of the dependent variable. Negative estimates mean a lower category of the dependent variable. The estimated value of -0.343 indicates that young customers have a worthy account.

Further the p-values of the Wald test indicate whether there are significant effects of the independent variables onto the dependent variable:

Parameter Estimates

		Estimate	Std. Error	Wald	df	Sig.	95% Confidence Interval	
							Lower Bound	Upper Bound
Threshold	[chist_class = 1,00]	-4,104	,241	290,378	1	,000	-4,576	-3,632
	[chist_class = 2,00]	-1,496	,212	49,683	1	,000	-1,912	-1,080
Location	[numcred_class=1,00]	-1,823	,117	242,468	1	,000	-2,052	-1,593
	[numcred_class=2,00]	0^a	.	.	0	.	.	.
	[othnstal=1,00]	-,456	,129	12,460	1	,000	-,709	-,203
	[othnstal=2,00]	-,318	,212	2,238	1	,135	-,734	,099
	[othnstal=3,00]	0^a	.	.	0	.	.	.
	[housng=1,00]	-,153	,177	,748	1	,387	-,500	,194
	[housng=2,00]	,069	,149	,213	1	,644	-,223	,361
	[housng=3,00]	0^a	.	.	0	.	.	.
	[age_class=1,00]	-,343	,156	4,843	1	,028	-,649	-,038
	[age_class=2,00]	-,022	,152	,022	1	,883	-,320	,275
	[age_class=3,00]	0^a	.	.	0	.	.	.
	[duration_class=1,00]	-,043	,108	,161	1	,689	-,254	,168
	[duration_class=2,00]	0^a	.	.	0	.	.	.

Link function: Complementary Log-log.
a. This parameter is set to zero because it is redundant.

Forecasting

The creditworthiness of a new customer should be predicted. The 34-years old new customer is living in his own house and has one installment debt to a bank with a duration time of eight months.

Enter the following values in row 1001 of the Data-View: numcred_class = 1

(X_1 classed), othnstal = 1 (X_2), housing = 2 (X_3), age_class = 2 (X_4 classed). Run one more time an ordinal regression. The predicted account-status is category 2 of the classed variable chist_class; i.e. the account-status of the new customer is neither good nor bad. The predicted category is listed in the 1001st row in the column PRE_2 of the Data-View.

8.5 Summary

If you want to run a regression analysis you have to select the correct procedure due to the level of the dependent variable as follows:

Regression Analysis		
	Level	
Name	Dependent Variable	Independent Variable(s)
Linear Regression	scale	scale
Binary logistic Regression	binary	nominal, ordinal, scale
Multinomial logistic Regression	nominal	nominal, ordinal
Ordinal Regression	ordinal	nominal, ordinal

9 Graphical Summaries of Data

In this chapter we will get to know some graphics that can be constructed with SPSS. You find an overview of all graphical options in SPSS here:

> Graphs → Chart Builder ... → Gallery

Or:

> Graphs → Legacy Dialogs

9.1 Scatter Plot

Before running the regression, you should visually check the scatter plot to determine whether a linear model is reasonable for these variables.

Example 9.1 (*vw_golf.sav*)
The price and the number of kilometers (English: kilometres, American: kilometers) covered of ten VW-Golfs are displayed in an insert of a newspaper. We want to construct a scatter plot of the independent variable "Kilometers" and the dependent variable "Price" with added regression line. The added fit line is $f(x) = 9632.299 - 0.04 \cdot x$. The coefficient of determination, $B = 0.52$, is the squared value of the correlation coefficient. It shows that about half the variation in price is explained by the model. The correlation coefficient is $r = -\sqrt{0.52} = -0.72$, i.e. there is a negative, but not strong, association between price and kilometers. High prices occur with low numbers of kilometers covered.

Input

Type of car	Year of first registration	km	Price (in €)
VW Golf	2016	96 800	7 450
VW Golf	2015	179 000	3 700
VW Golf	2014	138 000	3 490
VW Golf	2013	93 094	2 900
VW Golf	2008	196 891	1 190
VW Golf	2018	99 300	7 650
VW Golf	2016	67 000	6 300
VW Golf	2015	94 000	6 990
VW Golf	2015	126 000	3 600
VW Golf	2017	112 000	5 900

Commands

1) Graphs → Legacy Dialogs → Interactive → Scatterplot

2) Assign Variables
   X Axis = Number of km
   Y Axis = Price

3) Fit
   Method=Regression

4) OK

Output

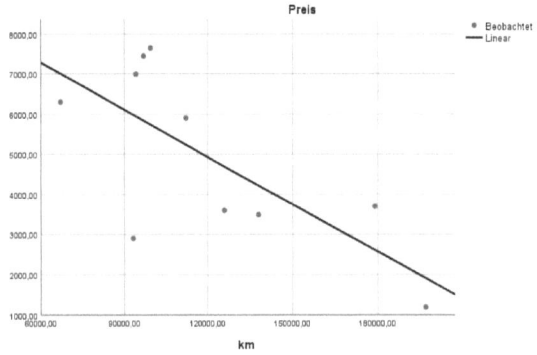

In the following example 9.2 we will construct a scatter plot with label cases.

154

Example 9.2 (*sales_profit_2016.sav* Source: Forbes)
In the year 2016 the top twenty companies of the world have the following sales and profits Mio. US$):

Rank Company	Sales	Profits	Rank
ICBC	171.1	44.2	1
China Construction Bank	146.8	36.4	2
Agricultural Bank of China	131.9	28.8	3
Berkshire Hathawy	210.8	24.1	4
JPMorgan Chase	99.9	23.5	5
Bank of China	122.0	27.2	6
Wells Fargo	91.4	22.7	7
Apple	233.3	53.7	8
Exxon Mobil	236.8	16.2	9
Toyota Motor	235.8	19.3	10
Bank of America	91.5	15.8	11
AT&T	146.8	13.2	12
Citigroup	85.9	15.8	13
HSBC	70.3	13.5	14
Verizon Communications	131.8	18.0	15
Wal-Mart Stores	482.1	14.7	16
Petro China	274.6	5.7	17
China Mobile	107.8	17.1	18
Samsung Electronics	177.3	16.5	19
Ping An Insurance	98.7	8.7	20
Allianz SE	115.4	7.3	21
Volkswagen AG	246.2	7.1	22
Microsoft	86.6	10.2	23
BNP Paribas	74.9	7.4	24
Daimler	165.7	9.3	25

Commands

1) Graphs → Chart Builder

2) ok

3) Select the first Scatterplot

4) Assign Variables
    X Axis =   Sales
    Y Axis =   Profits

5) Groups/Point-ID → Point-ID label

6) Point Label Variable="Corporation"

7) OK

Output

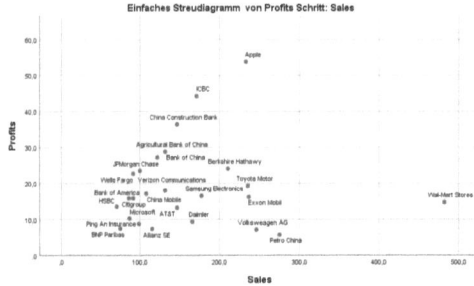

9.2   Bar Charts

Example 9.3 (*Redezeit_Oscar_Filme.sav* c.f. SZ 02/02/2019)
The so called Bechdel-Test (named after the US-American cartoonist Alison Bechdel) is a measure for the representation of women in fiction. It asks whether: Are there at least two women named in the fiction? Do they talk to each other? Do they talk to each other about something other than men? The Oscars' best Picture-winning films of the years 1991 until 2016 are analysed due to the proportion (in percent) of words spoken by characters with more than 100 words. The result is:

Year	Film	Men	Women
1991	Dances with Wolves	90.00	10.00
1992	The Silence of the Lambs	61.00	39.00
1993	Unforgiven	92.00	8.00
1994	Schindler's Ark	100.00	0.00
1995	Forrest Gump	83.00	17.00
1996	Braveheart	83.00	17.00
1997	The English Patient	76.00	24.00
1998	Titanic	63.00	37.00
1999	Shakespeare in Love	75.00	25.00
2000	American Beauty	59.00	41.00
2001	Gladiator	85.00	15.00
2002	A Beautiful Mind	81.00	19.00
2003	(Silent movie)		
2004	The Return of the King	96.00	4.00
2005	Million Dollar Baby	81.00	19.00
2006	Crash	76.00	24.00
2007	The Departed	92.00	8.00
2008	No Country For Old Men	85.00	15.00
2009	Slumdog Millionaire	93.00	7.00
2010	The Hurt Locker	100.00	.00
2011	The King's Speech	90.00	10.00
2012	(Musical film)		
2013	Argo	96.00	4.00
2014	12 Years a Slave	82.00	18.00
2015	Birdman	69.00	31.00
2016	Spotlight	94.00	6.00

1) Graphs → Legacy Dialogs → Bar …

2) Stacked
   Data in Chart Are "Summaries of separate variables"
   Define

3) Bars Represent = Proportion women
   = Proportion men

(Change Statistic = Sum of values)
Category Axis = Tiltle

4) OK

Output

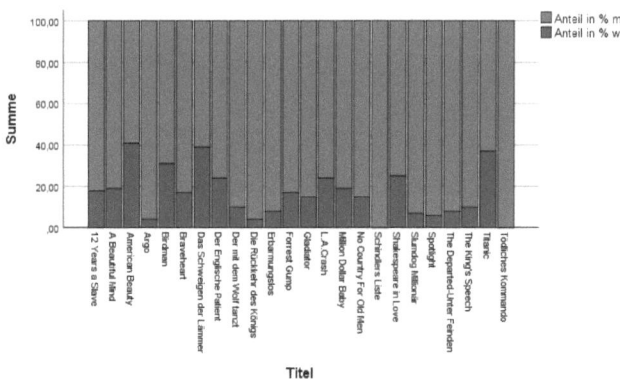

The black-and-white diagram shows how little women speak in Oscar-wining movies: on average 16.6%.

9.3 Pie Chart

Example 9.4 (*Im_Ex_2005.sav* Source: WTO)
We want to show a pie chart with imports (in US$) and a pie chart with exports (in US$) of every region Africa, Asia, Commonwealth and Independent States, Europe, Middle East, North America, South and Central America:

Region	Population 2005	Imports 2005	Exports 2005
Africa	915 210 928	249 259	593 583
Asia	3 667 774 066	2 987 616	6 589 809
CIS	33 956 977	215 958	675 659
Europe	807 289 020	4 440 739	3 987 984
Middle East	190 084 161	322 136	1 069 630
North America	331 473 276	2 283 704	1 477 493
South and Central America	553 908 632	295 319	354 573
Total	6 499 697 060	10 794 730	14 748 730

Population: *www.internetworldstats.com*
Exports/Imports: *www.wto.org*

Commands

1) Graphs → Legacy Dialogs → Pie ...

2) Data in Chart Are "Values of individual cases"
   Define

3) Slices Represent = I_Year_2005
   Slice Labels, Variable = Region

4) ok

Output

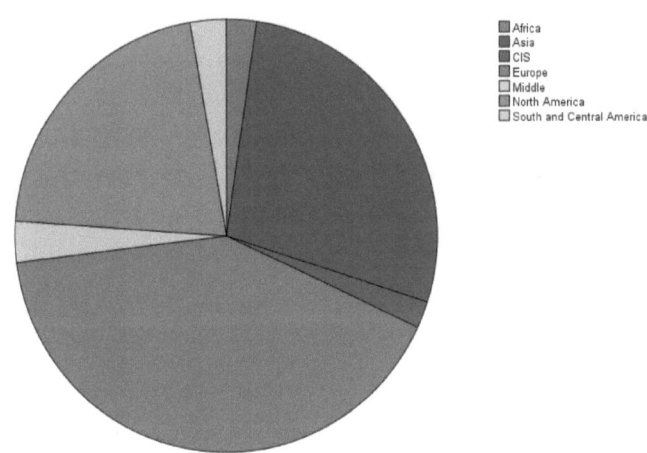

9.4 Boxplot (Whisker-Plot)

In practice, often one would like to have a quick visual comparison for two data sets. For this purpose, the following five key figures should be calculated per data record:

1) The smallest value $x_{\min}$.

2) The 25%-point $x_{0.25}$.

3) The median (50%-point) $x_{0.50}$.

4) The 75%-point $x_{0.75}$.

5) The largest value $x_{\max}$.

A **boxplot** (or whisker-plot) is a diagram of a sample through its five-number summaries: the smallest observation (sample minimum), lower quartile (25%-point), median (50%-point), upper quartile (75%-point), and largest observation (sample maximum). The spacings between the different parts of the box help indicate the degree of dispersion (spread) and skewness in the data. Boxplots can be drawn either horizontally or vertically:

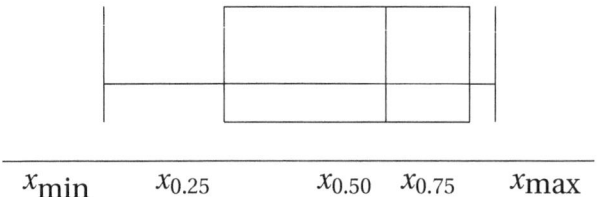

$$x_{\min} \qquad x_{0.25} \qquad\qquad x_{0.50} \quad x_{0.75} \qquad x_{\max}$$

Remark: The US-American StatisticanTukey (1915 - 2000) suggested a data point as an x **outlier**, if x is lying more than 1.5-times of the quartile distance below the lower quartile or above the upper quartile:

$$x=\text{Outlier} \Leftrightarrow x < x_{0.25} - 1.5 \cdot (x_{0.75} - x_{0.25})$$
$$x=\text{Outlier} \Leftrightarrow x > x_{0.75} + 1.5 \cdot (x_{0.75} - x_{0.25})$$

Example 9.5 (*G8_Exports_1995_2017.sav* Source: WTO)

In the years 1995 to 2017 exports (in Mio. US$) of the G8 countries US, Canada, Germany, Italy, France, United Kingdom, Japan, Russian Federation amounted:

Year	CDN	USA	F	D	I	GB	RUS	CHN	JP
1995	192197	584743	301162	523461	233766	237953	81095	148780	443116
1996	201633	625073	305509	524649	252293	258527	88600	151048	410901
1997	214422	689182	302144	512891	240414	280406	88330	182792	420957
1998	214327	682138	320631	543752	245801	273949	74884	183712	387927
1999	238446	695797	325520	543529	235559	272161	75665	194931	417610
2000	276635	781918	327616	551818	240521	285429	105565	249203	479249
2001	259858	729100	323379	571645	244490	272715	101884	266098	403496
2002	252394	693103	331719	615831	254427	280195	107301	325596	416726
2003	272739	724771	392039	751560	299333	305627	135929	438228	471817
2004	316547	818775	452106	909887	353782	347493	183207	593326	565675
2005	359399	904383	460157	969858	367200	382761	243569	761954	594905
2006	388178	1025967	495868	1108107	416875	448653	303551	968980	646725
2007	420693	1148199	559624	1321352	500088	439109	354403	1220450	714327
2008	456471	1287442	615913	1446392	543050	459685	471606	1430690	782047
2009	316713	1056043	484725	1126383	405777	352491	303388	1201610	580719
2010	388019	1278263	520661	1268874	447535	405666	400132	1577820	769839
2011	452000	1480000	596000	1472000	522000	473000	522000	1906312	823000
2012	454840	1547280	569070	1407100	500240	468370	529000	2048710	798570
2013	458000	1580000	580000	1453000	518000	542000	523000	2209010	715000
2014	475000	1621000	583000	1508000	529000	506000	498000	2342290	684000
2015	408000	1505000	506000	1329000	459000	460000	340000	2273470	625000
2016	390120	1454610	501260	1339650	461520	412100	259300	2098160	644930
2017	433000	1576000	541300	1401000	499100	436500	336800	2157000	683300

Commands

1) Open the file *G8_Exports_1995_2017.sav*

2) Graphs → Legacy Dialogs → Boxplot

3) Select "Simple"
   Select "Summaries of separate variables"
   Click "Define"

4) Boxes Represent: US, Canada, Germany, Italy, France, United Kingdom, Japan, Russian Federation
   Label Cases by = Year

5) ok

Output

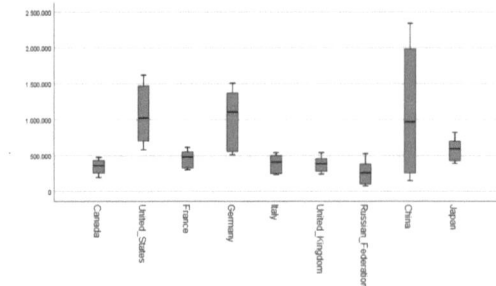

The "belt" (median) of Germany is the highest, i.e. between 1995 and 2017, Germany has the highest median export volume (measured in millions of US$), the lowest being the Russian Federation, i.e. Russia has the smallest median export volume.

Furthermore, the interquartile range, i.e. the distance between the 25% and 75% points, is the largest for China and the smallest for United Kingdom, i.e. in China, annual exports are the most volatile and in Great Britain, annual exports do not fluctuate much.

Example 9.6 (*survey_sample.sav* Source: Tutorial von IBM)
In the example *survey_sample.sav* of the SPSS-Tutorial we get the following boxplot of the tv consumption separated by women and men:

Commands

1) Open the file *survey_sample.sav.*

2) Graphs → Chart Builder → ok

3) Boxplot

4) *x*-Axis = sex
   y-Axis = tv hours

5) ok

Output

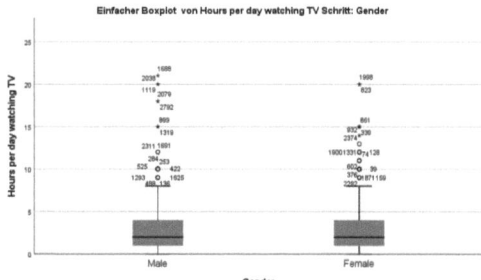

Einfacher Boxplot von Hours per day watching TV Schritt: Gender

In the box plot the two medians are at the same height, also the distances between the 25% point and the 75% point are the same. So there are no differences in the TV consumption of women and men.

SPSS classifies the many isolated points in the boxplot as outliers.

10 Analysis of Variance

Main purpose: The population is divided into three or more different groups. We want to test whether the distribution of one variable differs between this groups. The random variable must be leveled as follows:

nominal:	no
ordinal:	no
scale:	yes

The analysis of variance has the following assumptions: Normal distribution of the variable in each group, the variable must be stochastically independent across the groups and the variances of the variable in each group must be the same (homogeneous variances).

For small sample sizes we check the Normal distribution with a goodness-of-fit test (Lilliefors, Shapiro-Wilk, Jarque-Bera). For large samples sizes we check the Normal distribution with a QQ plot. The homogeneity of the variances can be verified with the Levene test.

10.1 Three or More Groups

Example 10.1 (*Aisle_Location.sav* c.f. Berenson et al., page 489)
The retailing manager of a supermarket chain wants to determine whether product location has any effect on the sales of pet toys. Three different aisle locations are considered: front, middle, and rear.

    1. Group: Sales of stores with the product in the front.

    2. Group: Sales of stores with the product in the middle.

    3. Group: Sales of stores with the product in the rear.

Question: Does the variable "Sales" (in 1 000 US $) have the same distribution in each group?

The test statistic $F_{emp.}$ is the ratio of two variances, therefore this test is called "$\boxed{an}$ $alysis$ $\boxed{o}$f $\boxed{va}$ riance" (abbreviation: ANOVA).

A random sample of 18 stores is selected, with six stores randomly assigned to each aisle location. The size of the display area and price of the product are constant in all stores. And the end of a one-month trial period, the sales volumes (in thousands of dollars) of the product in each store were as follows:

Sales (in 1 000 US $)

Location of the Product		
front	middle	rear
8.6	3.2	4.6
7.2	2.4	6.0
5.4	2.0	4.0
6.2	1.4	2.8
5.0	1.8	2.2
4.0	1.6	2.8

$Y_{ij}=$ Sales in the i-th group of the j-th store

The assumptions of ANOVA are: Normal distribution of the variable Y sales in all three groups, equal variances σ^2 in the three groups and the three variables must be stochastically independent.

First, we check the Normal distribution of the sales: For the location "front" we get the skewness $S_{SPSS} = 0.495$, i.e. $S = .452$ and the kurtosis $K_{SPSS} = -0.306$, i.e. $K = 2.038$; the p-value of the Shapiro-Wilk test is 0.948. For the location "middle" we get the skewness $S_{SPSS} = 1.158$, i.e. $S = 1.057$ and the kurtosis $K_{SPSS} = 1.103$, i.e. $K = 2.521$; the p-value of the Shapiro-Wilk test is 0.492. For the location "rear" we get the skewness $S_{SPSS} = 0.745$, i.e. $S = 0.680$ and the kurtosis $K_{SPSS} = -0.361$, i.e. $K = 2.019$; the p-value of the Shapiro-Wilk test is 0.565. This means that the variable sales is Normally distributed within each group.

Secondly, we check homogeneity of the variances with the Levene test. The p-value of the Levene test is 0.142; this means homogeneity of the variances.

Thirdly, we have no multivariate data set. Therefore we cannot check stochastic independence with SPSS. But we assume that the three variables $Y_1 =$ "Sales (in 1 000 US $ per month) of the product located in the front", $Y_2 =$ "Sales (in 1 000 US $ per month) of the product located in the middle", $Y_3 =$ "Sales (in 1 000 US $ per month) of the product located in the rear" are stochastically independent. (In this example we consider the special case of equal sample sizes $n_1 = n_2 = n_3 = 6$ within each group.)

Having Normal distribution with equal variances the sales can only differ in their mean values μ_1, μ_2, μ_3:

$$Y_1 \sim N(\mu_1; \sigma^2)$$
$$Y_2 \sim N(\mu_2; \sigma^2)$$
$$Y_3 \sim N(\mu_3; \sigma^2)$$

With the ANOVA we check the following testing problem at level $\alpha = 0.05$:

H_0: $\mu_1 = \mu_2 = \mu_3$
versus
H_1: $\mu_i \neq \mu_k$ for at least one index pair $(i; k)$

Expressed in words:

Analysis of Variance
H_0 : There is no difference among the means; brief: $\mu_1 = \mu_2 = \mu_3$
versus
H_1 : At least two means differ; brief: $\mu_i \neq \mu_k$ for at least one index pair $(i; k)$
Rejection of $H_0 \Leftrightarrow p$-value ≤ 0.05

The empirical value of F is:

$$F_{emp.} = \frac{\frac{1}{I-1}\sum_{i=1}^{I} n_i(\bar{y}_i - \bar{y})^2}{\frac{1}{n-I}\sum_{i=1}^{I}\sum_{j=1}^{n_i}(y_{ij} - \bar{y}_i)^2} = \frac{\frac{1}{2} \cdot 14.444}{\frac{1}{15} \cdot 25.760} = \frac{24.222}{1.7173} = 14.105$$

The p-value of the ANOVA is:

$$P_{I-1;n-I}(F > F_{emp.}) = P_{2;15}(F > 14.105) = 0.0003579336$$
$$p\text{-value} \approx 0.0004 < 0.05 = \alpha$$

This means rejection of H_0; this means that the mean values of the sales in the three groups differ significantly.

Interpretation: The average sales of stores with the product located in the front is 6 067 $US per month. The average sales of stores with the product located in the middle is 2 067 $US per month. The average sales of stores with the product located in the rear is 3 733 $US per month. In the sample we have different average values of the sales. And these differences in the sample enable us to say that we have significant differences in the population, too. Because the sample is a representative sample.

10.2 Special Case: Two Groups

In the case of only two groups we run the t-test or the Welch test instead of the ANOVA depending on whether the two variances are the same.

Example 10.2 (*Pisa-Studie-Gruppen.sav* c.f. Handl [2002] p. 294)
All participant countries of the PISA-Program (Program for International Student Assessment) in the year 2000 are split into three groups:

    1. Group: All countries with low expenditure of time for home-work

    2. Group: All countries with moderate expenditure of time for homework

3. Group: All countries with high expenditure of time for homework

Question: Does the variable "mathematical literacy" (in points) have the same distribution within the first and the third group?

The following table shows the points for mathematical literacy of every participant country:

1. Group		3. Group	
Country	Points	Country	Points
FIN	536	GR	447
J	557	GB	529
FL	514	IRL	503
L	446	I	457
A	515	LV	463
S	510	MEX	387
CH	529	PL	470
CZ	498	RUS	478
		E	476
		H	488
$n_1 = 8$ $\bar{y}_1 = 513.125$		$n_3 = 10$ $\bar{y}_3 = 469.8$	
$n = n_1 + n_3 = 18$ and $\bar{y}_1 - \bar{y}_3 = 43.325$			

$Y_{ij} =$ Points for mathematical literacy in the i-th Group of the j-th country

We have to check the Normal distribution of the variable Y mathematical literacy in the first and in the third group and we have to check the homogeneity of the variances σ^2.

For the first group we get the skewness $S_{SPSS} = -1.128$, i.e. $S = -1.055$ and the kurtosis $K_{SPSS} = 2.487$, i.e. $K = 3.518$; the value of the Jarque-Bera test statistic is $T = 1.573$. For the third group we get the skewness $S_{SPSS} = -0.875$, i.e. $S = -0.830$ and the kurtosis $K_{SPSS} = 2.412$, i.e. $K = 3.861$; the value of the Jarque-Bera test statistic is $T = 1.457$. Therefore the variable literacy has Normal distribution in both groups.

We check the homogeneity of the variances with the Levene test. We get the p-value 0.761; this means homogeneity of the variances.

It is not possible to check the stochastic independence with SPSS, because we have no bivariate data set but two univariate data sets. Therefore we assume that the two variables Y_1 = "Points for mathematical literacy in group 1", Y_3 = "Points for mathematical literacy in group 3"' are stochastically independent.

Due to the Levene test and due to the Shapiro-Wilk test we have homogeneity of the variances and Normal distribution. Perhaps the two mean values μ_1, μ_3 may differ:

$$Y_1 \sim N(\mu_1; \sigma^2)$$
$$Y_3 \sim N(\mu_3; \sigma^2)$$

We consider the t-test for independent samples at level $\alpha = 0.05$:

$$H_0 : \mu_1 = \mu_3 \text{ versus } H_1 : \mu_1 \neq \mu_3$$

The empirical value of the t-statistic is:

$$t_{emp.} = \frac{\overline{y}_1 - \overline{y}_2}{\sqrt{\frac{1}{n_1+n_2-2} \sum_{i=2}^{2} \sum_{j=1}^{n_i} (y_{ij} - \overline{y}_i)^2}} = 2.578$$

We get the following p-value of the t-test:

$$
\begin{aligned}
2 \cdot P_{n_1+n_2-2}(t > | \, t_{emp.} \, |) &= 2 \cdot P_{16}(t > | \, 2.578 \, |) \\
&= 2 \cdot P_{16}(t > 2.578) \\
&= 2 \cdot 0.01011188 \\
&= 0.020
\end{aligned}
$$

Further: p-value $= 0.020 \leq 0.05 = \alpha$

This means rejection of H_0; this means that the mean value of the mathematical literacy differs significantly for pupils with less and pupils with time-consuming homework.

Remark: In case of two groups the p-value of the t-test and the p-value of the ANOVA are the same.

10.3   Summary

The analysis of variance is a test that verifies whether the mean values of a variable in different groups are the same, if Normal distribution, stochastic independence and homogeneity of the theoretical variances may be assumed.

10.4   SPSS-Commands

10.4.1   Compare Two Means

1) Open the file "Pisa-Studie-Gruppen.sav"

2) Analyze $\rightarrow$ Compare Means $\rightarrow$ Independent-Samples t-Test

3) Test Variable = "Math.Grundbildung"
   Grouping Variable="Group"
   Click "Define Groups ... "
   Group 1 = 1
   Group 2 = 3
   Click "Continue".

4) Click "Options ... ", if you want to change the confidence level.
   Confidence Interval Percentage = 95 %.
   Click "Continue".

5) Click "OK".

The p-value of the Levene test is 0.761. This means that the variances are homogeneous.

The p-value of the t-test is 0.020. This means that the two means differ significantly.

Output

SPSS Output

Independent Samples Test

		Levene's Test for the Equality of Variances		t-test for Equality of Means					95% Confidence Interval of the Difference	
		F	Sig.	t	df	Sig. (2-tailed)	Mean Difference	Std. Error Difference	Lower	Upper
Math.Grund	Equal variances assumed	.096	.761	2.578	16	.020	43.325	16.803	7.703	78.947
	Equal variance not assumed			2.620	15.845	.019	43.325	16.533	8.248	78.402

172

10.4.2 Compare Three or More Means

1) Open the file "Aisle_Location.sav"

2) Analyze → Compare Means → One-Way ANOVA

3) Dependent List = "Sales"
   Factor="Location"

4) Click "Post Hoc ..." and select "R-E-G-W F". The test level is 0.05.
   Click "Continue".

5) Click "Options" and select under Statistics "Descriptive" and "Homogeneity of variance test".
   (If we select "Means plot" we get a plot with the line that connects the
   three average scores in mathematical literacy.)
   Click "Continue".

6) Click "OK".
   The p-value 0.000 is in the column "Significance" of the table "ONEWAY
   ANOVA".

Output
Test of the Homogeneity of the Variances (Based on Mean)
Sales

Levene-Statistic	df1	df2	Significance
2.227	2	15	.142

The p-value of the Levene test (Based on Mean) is 0.142. This means that
we do not reject the null hypothesis of equal variances. Please remember
that homogeneity of the variances is an important assumption for the analysis of variance. (If we have heterogeneity of the variances we cannot run
the analysis of variance.)

ONEWAY ANOVA

Sales

	Sum of Squares	df	Mean Square	F	Sig.
Between Groups	48.444	2	24.222	14.105	.000
Within Groups	25.760	15	1.717		
Total	74.204	17			

The *p*-value of the ANOVA is 0.000; this means that we reject the null hypothesis; this means that the mean sales differ significantly between the three groups.

There will be significant differences in the mean values, if the *p*-value of the ANOVA is less or equal 0.05. It can be determined with Ryan-Einot-Gabriel-Welsch test (brief: R-E-G-W), which group means are significantly different. The findings of the R-E-G-W test are listed in a table. The mean values that differ significantly are marked.

11 Kruskal-Wallis Test

Main purpose: The population is split into two or more different groups. We want to check whether the location of the distribution of one variable differs between this groups. The random variable must be leveled as follows:

nominal:	no
ordinal:	yes
scale:	yes

The so called Kruskal-Wallis test is a nonparametric alternative to the A-NOVA in case that the assumption of Normal distribution is violated but stochastic independence is given, i.e. the samples are stochastically independent. The Kruskal-Wallis test compares the fifty percent points of a special variable in each of several groups:

Kruskal-Wallis test
H_0:   The theoretical medians are the same in each group.
versus
H_1:   At least two theoretical medians differ.
Rejection of H_0 ⇔ p-value $\leq \alpha$

Example 11.1 (*Williams.sav* c.f. Andersonet al. p. 740)
Williams Manufacturing Company Limited hires employees for its management staff from three local colleagues A, B, C. Recently, the company's personnel department began collecting and reviewing X="annual performance ratings of a manager (scaled 0 - 100)" in an attempt to determine whether there are differences in performance among the managers hired from these colleagues. In words:

H_0: The theoretical medians of the performance are the same across all three colleague-groups.

H_1: At least two theoretical medians differ.

We assume that the performance ratings are stochastically independent.

Performance rating data are available from independent samples of $n_1 = 7$ employees from colleague A, $n_2 = 6$ employees from colleague B and $n_3 = 7$ employees from colleague C:

Perfor-mance	Col-league	Perfor-mance	Col-league	Perfor-mance	Col-league
25	A	60	B	50	C
70	A	20	B	70	C
60	A	30	B	60	C
85	A	15	B	80	C
95	A	40	B	90	C
90	A	35	B	70	C
80	A			75	C

The midranks of the $n = n_1 + n_2 + n_3 = 20$ observations are:

Perfor-mance	Rank	Perfor-mance	Rank	Perfor-mance	Rank
25	3	60	9	50	7
70	12	20	2	70	12
60	9	30	4	60	9
85	17	15	1	80	15,5
95	20	40	6	90	18,5
90	18,5	35	5	70	12
80	15,5			75	14
Σ	$R_1 = 95$	Σ	$R_2 = 27$	Σ	$R_3 = 88$

Mid ranks are used for tied sample values. For example the rating 60 points is observed three times. If these three values differ slightly the ranks 8, 9, 10 would be the corresponding ranks of these values. The arithmetic mean $\frac{1}{3}(8 + 9 + 10) = 9$ of these three ranks is the mid rank 9.

176

In case of no ties the test statistic H of the Kruskal-Wallis test is as follows:

$$H = \frac{12}{n \cdot (n+1)} \sum_{i=1}^{k} \frac{R_i^2}{n_i} - 3(n+1)$$

where $R_1, R_2, \ldots R_k$ are denoting the rank sums in every of the k samples.

The considered sample has ties. The seven smallest ranks are only available once. The ranks 9 and 12 occur three times respectively. The rank 14 is only available once, the mid rank 15.5 two times, the rank 17 once, the mid rank 18.5 two times and the rank 20 once. The 14-tuple t shows how often each of the 14 different mid ranks occur: $t = (1,1,1,1,1,1,1,3,3,1,2,1,2,1)$. The sum $\sum_{j=1}^{14} (t_j^3 - t_j)$ equals $2 \cdot (3^3 - 3) + 2 \cdot (2^3 - 2) = 60$, because: $1^3 - 1 = 0$. The factor for the ties correction is:

$$1 - \frac{1}{n^3 - n} \cdot \sum_{j=1}^{14} \left(t_j^3 - t_j \right) = 1 - \frac{1}{20^3 - 20} \cdot 60 = 0.9924812$$

(In case of not tied values the factor for the correction is evaluated as one.) The test statistic H of the Kruskal-Wallis test is divided by the correction factor:

$$\frac{H}{0.9924812} = \frac{\frac{12}{20 \cdot 21} \left(\frac{95^2}{7} + \frac{27^2}{6} + \frac{88^2}{7} \right) - 3 \cdot 21}{0.9924812} = \frac{8.916327}{0.9924812} = 8.9839$$

The p-value of the Kruskal-Wallis test is evaluated approximatively from a chi-square-distribution with $df = k - 1 = 2$:

$$p\text{-value} \approx P_{df=2}(H > 8.9839) = 0.01119893 \approx 0.011.$$

At least two theoretical medians differ significantly. We conclude that the three populations are not identical. Manager performance differs significantly depending on the colleague attended.

Remark: In this example the analysis of variance could also have been carried out. The Shapiro-Wilk test verifies Normal distribution of the performance in each group with the p-values 0.197 resp. 0.772 resp. 0.974. The p-value 0.398 of the Levene-test indicates homogeneity of the variances. Finally the p-value 0.002 of the ANOVA provides significant differences in the mean values of the performance within the three colleague-groups.

⚠ The ANOVA has a higher asymptotic relative efficiency than the Kruskal-Wallis test (c.f. Gibbons [2010]). If the requirements for an ANOVA are met, the ANOVA is preferable to the Kruskal-Wallis test.

11.1 Summary

The Kruskal-Wallis test verifies whether two or more theoretical medians are the same in case of no specified distribution.

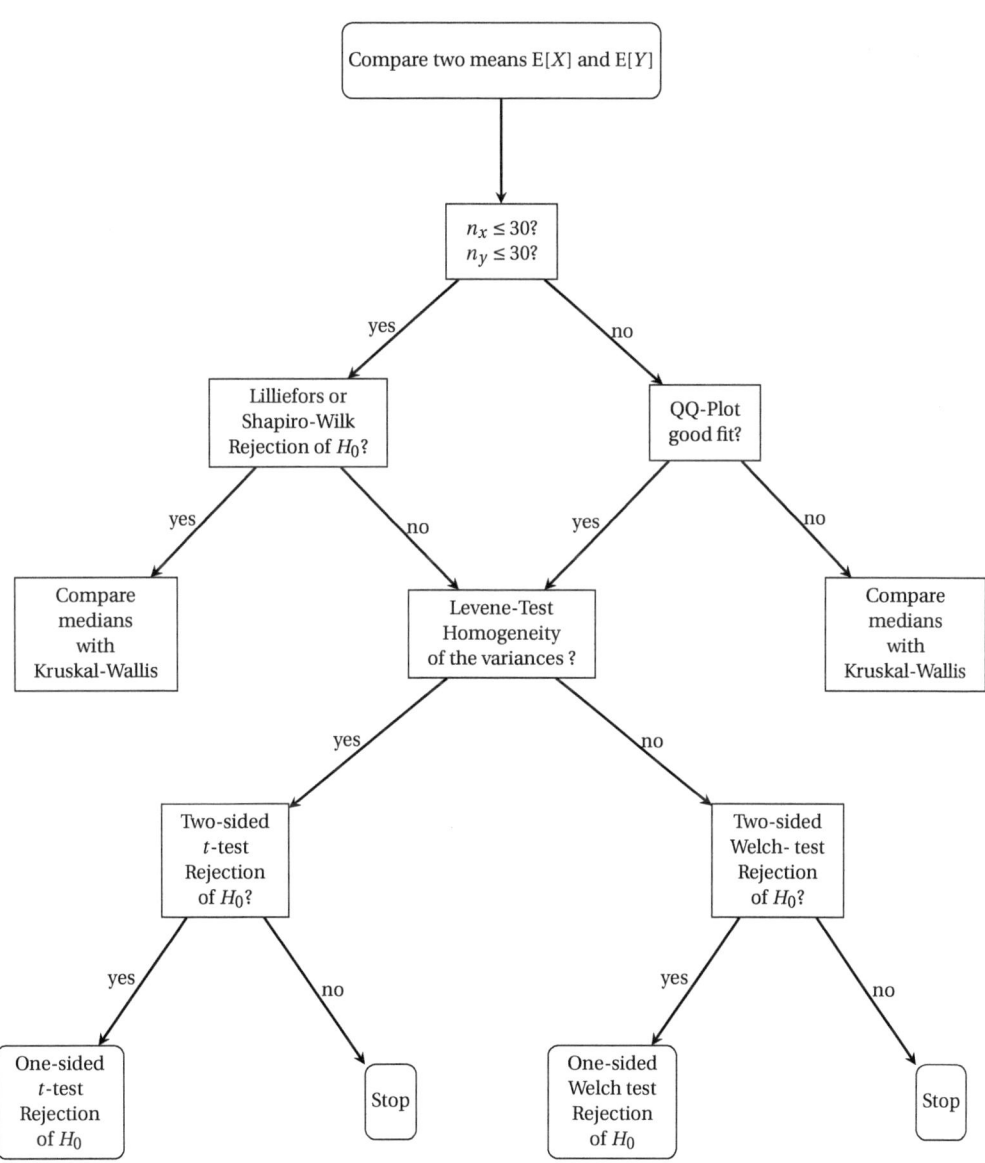

11.2 SPSS-Commands

First the string variable Colleague must be recoded into a numeric variable C_num with $A=1$, $B=2$, $C=3$.

Commands

1) Open the file "Williams.sav".

2) Analyze → Nonparametric Tests → Legacy Dialogs → K Independent Samples ...

3) Test Variable List = "'Performance"'
   Grouping Variable = "'C_num"'
   Click "Define Range ... ".
   Minimum = 1
   Maximum = 3
   Continue

4) ok

The p-value 0.011 is listed in the row "Asymptotic Significance".

Output

Test Statistics[a,b]

	Performance
Kruskal-Wallis H	8.984
df	2
Asymp. Sig.	.011

a. Kruskal Wallis Test
b. Grouping Variable: C_num

12 Principal Components Analysis

Purpose: A data set of observations of several variables is converted into a set of only a few variables called principal components. The number of principal components is less than or equal to the number of original variables. The principal components are linear combination of the observed variables. The variables must be leveled as follows:

nominal:	no
ordinal:	no
scale:	yes

We want to use the principal components analysis to get an ordering of the cases/objects based on the values of the first principal component or to get a scatter plot of all cases/objects based on the values of the first and second principal components as x- and y-axes.

12.1 Univariate Data Set

If we have a data set of observed values of only one scale leveled variable, for example X=age, the objects may be ordered descending or ascending. Therefore a principal components analysis is obsolete.

12.2 Bivariate Data Set

The values of a bivariate data set may be plotted in a scatter plot without any problem. But how to order the n cases/objects?

Example 12.1

We want to order five countries of the PISA survey 2000 based on the reading literacy and the mathematic literacy. The observed points of literacies are as follows:

Country	Reading	Mathematics
Denmark	497	514
Greece	474	447
Italy	487	457
Portugal	470	454
Sweden	516	510

We want to extract one principal component to order the five countries.

The arithmetic means are:

488.8 points   Reading
476.4 points   Mathematics

We construct a scatter plot:

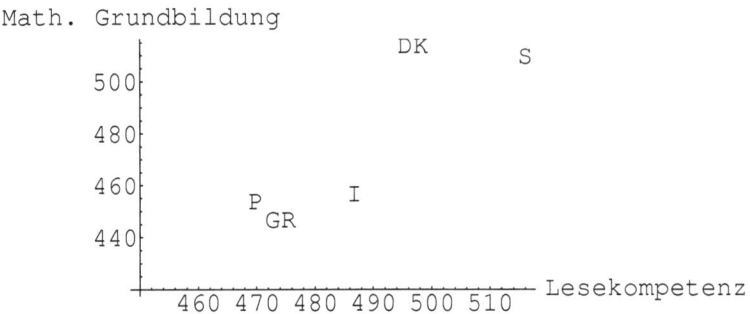

The deviation of mathematics is larger than the deviation of reading. The standard deviations are:

$$\sqrt{345.7} = 18.6 \text{ points} \quad \text{Reading}$$
$$\sqrt{1\,071.3} = 32.7 \text{ points} \quad \text{Mathematics}$$

The first principal component should represent the values of the two variables. We standardize the observed values (value minus arithmetic mean divided by the standard deviation):

$$\frac{\text{Reading Lit.} - 488.8}{18.6} \quad \text{resp.} \quad \frac{\text{Math. Lit.} - 476.4}{32.7}$$

The standardized values are:

Country	Reading Standardized value x_i	Mathematics Standardized value y_i
Denmark	0.441	1.149
Greece	−0.796	−0.898
Italy	−0.097	−0.593
Portugal	−1.011	−0.684
Sweden	1.463	1.027

The arithmetic means $\overline{x}$ and $\overline{y}$ of the standardized values are zero and the standard deviations s_x and s_y are one.

Based on these values x_i, y_i we construct a linear combination:

$$b_1 \cdot x_i + b_2 \cdot y_i \; ; b_1, b_2 \in \mathbb{R}$$

The linear combination should have a very large deviation for a better discrimination. The deviation of the linear combination will increase if the values b_1, b_2 increase, so we have to give un upper bound for b_1 and b_2:

$$b_1 + b_2 = 1$$

Resp. for a simple calculation:

$$b_1^2 + b_2^2 = 1$$

That is a mathematical optimization problem:

183

Deviation of $(b_1 x_i + b_2 y_i) \overset{!}{=} \max.$
under the restriction $b_1^2 + b_2^2 = 1$

The empirical variance of the values $(b_1 x_1 + b_2 y_1), (b_1 x_2 + b_2 y_2), \ldots, (b_1 x_n + b_2 y_n)$ is:

$$b_1^2 s_x^2 + b_2^2 s_y^2 + 2 b_1 b_2 s_{xy} = b_1^2 + b_2^2 + 2 b_1 b_2 s_{xy}$$

s_{xy} denotes the empirical covariance. Due to the method of substitution (c.f. Arrenberg "'Wirtschaftsmathematik für Bachelor"') we get the following optimization problem:

$$f(b_1, b_2) = b_1^2 + b_2^2 + 2 b_1 b_2 s_{xy} \overset{!}{=} \text{maximum}$$
$$\text{on the restriction: } b_1^2 + b_2^2 = 1 \Leftrightarrow b_2 = \pm \sqrt{1 - b_1^2}$$

We distinguish between two cases. In the first case that $b_2 = +\sqrt{1 - b_1^2}$ we get the following optimization function due to the method of substitution:

$$f(b_1) = b_1^2 + 1 - b_1^2 + 2 b_1 \sqrt{1 - b_1^2}\, s_{xy} = 1 + 2 b_1 \sqrt{1 - b_1^2}\, s_{xy}$$

Derivation of the first order:

$$f'(b_1) = 2\sqrt{1 - b_1^2}\, s_{xy} + 2 b_1 \cdot \frac{1}{2} \cdot \frac{-2 b_1}{\sqrt{1 - b_1^2}} \cdot s_{xy} = 2\sqrt{1 - b_1^2}\, s_{xy} - \frac{2 b_1^2}{\sqrt{1 - b_1^2}} \cdot s_{xy}$$

First order necessary condition:

$$0 = 2\sqrt{1 - b_1^2}\, s_{xy} - \frac{2 b_1^2}{\sqrt{1 - b_1^2}} \cdot s_{xy}$$

Division by $2 s_{xy}$:

$$0 = \sqrt{1 - b_1^2} - \frac{b_1^2}{\sqrt{1 - b_1^2}} = \frac{1 - b_1^2 - b_1^2}{\sqrt{1 - b_1^2}} = \frac{1 - 2 b_1^2}{\sqrt{1 - b_1^2}}$$

Multiplication with the factor $\sqrt{1 - b_1^2}$:

$$0 = 1 - 2b_1^2 \Leftrightarrow b_1^2 = \frac{1}{2}$$

For the sufficient condition we check whether the sign of the second order derivation is negative:

$$f''(b_1) = 2 \cdot \frac{1}{2} \cdot \frac{-2b_1}{\sqrt{1 - b_1^2}} \cdot S_{xy} - \frac{4b_1\sqrt{1 - b_1^2} - 2b_1^2 \cdot \frac{-2b_1}{\sqrt{1 - b_1^2}}}{1 - b_1^2} S_{xy}$$

Summarizing terms:

$$f''(b_1) = \frac{-2b_1}{\sqrt{1 - b_1^2}} S_{xy} - \frac{4b_1}{\sqrt{1 - b_1^2}} S_{xy} - \frac{4b_1^3}{\left(1 - b_1^2\right)^{1.5}} S_{xy}$$

We get:

$$f''(b_1) = -\frac{6b_1}{\sqrt{1 - b_1^2}} S_{xy} - \frac{4b_1^3}{\left(1 - b_1^2\right)^{1.5}} S_{xy}$$

For $b_1 = \frac{1}{\sqrt{2}}$ we get: $f''(b_1) = -8.681934$. And for $b_1 = -\frac{1}{\sqrt{2}}$ we get : $f''(b_1) = 8.681934$; i.e. $b_1 = \frac{1}{\sqrt{2}}$ is a local maximum.

For the first case of the case distinction we get: $b_2 = +\sqrt{1 - b_1^2} = \frac{1}{\sqrt{2}}$

Hence $f(b_1, b_2)$ has a local maximum in the point $(b_1, b_2) = \left(\frac{1}{\sqrt{2}}, \frac{1}{\sqrt{2}}\right)$ taking into account the restriction.

For the second case $b_2 = -\sqrt{1 - b_1^2}$ we get that $f(b_1, b_2)$ has a local maximum in the point $(b_1, b_2) = \left(-\frac{1}{\sqrt{2}}, -\frac{1}{\sqrt{2}}\right)$ taking into account the restriction. The interpretation regarding the first principal component is the same for both cases $b_1 = b_2 = 0.707$ and $b_1 = b_2 = -0.707$. Therefore we take as a solution of the principal components analysis:

$$b_1 = b_2 = \frac{1}{\sqrt{2}} = 0.707$$

SPSS makes a rotation and we get:

$$b_1 = 0.517$$
$$b_2 = 0.517$$

Therefore we obtain the linear combination:

$$0.517x_i + 0.517y_i$$

The values of the linear combination based on the standardized values of reading x_i and mathematics y_i are:

Country	Linear combination
Denmark	0.82246
Greece	−0.87649
Italy	−0.35672
Portugal	−0.87715
Sweden	1.28790

Countries with high scores in both reading and mathematics also have high scores on the first principal component. (If both scores of the first principal component are negative, for example 1st principal component $= -0.6x_i - 0.8y_i$, high-performing countries would have very small values with respect to the first principal component.)

If we order the five countries based on their values of the first principal component we get the following order: Sweden has the best literacy followed by Denmark and Italy. Greece and Portugal do poorly.

If we had just added the two points for reading and maths, the ranking would be: Sweden, Denmark, Italy, Portugal, Greece.

Generally the coefficients b_1 and b_2 of the linear combination of the first principal component based on a bivariate data set are always $b_1 = b_2 = 0.517$, if only one principal component is extracted.

12.3 Multivariate Data Set

Now we consider a principal components analysis based on a multivariate data set.

Example 12.2 (*Pisa-Studie-Hauptk.sav*)
The first PISA survey in the year 2000 involved 31 countries. The observed variables are

- Reading literacy (in points)

- Mathematical literacy (in points)

- Natural Science literacy (in points)

The observed values are:

Country	Reading	Mathematics	Science
Australia	528	533	528
Austria	507	515	519
Belgium	507	520	496
Brazil	396	334	375
Canada	534	533	529
Czech Republic	492	498	511
Denmark	497	514	481
Finland	546	536	538
France	505	517	500
Germany	484	490	487
Great Britain	523	529	532
Greece	474	447	461
Hungary	480	488	496
Ireland	527	503	513
Iceland	507	514	496
Italy	487	457	478
Japan	522	557	550
Latvia	458	463	460
Liechtenstein	483	514	476
Luxembourg	441	446	443
Mexico	422	387	422
New Zealand	529	537	528
Norway	505	499	500
Poland	479	470	483
Portugal	470	454	459
Russia	462	478	460
South Korea	525	547	552
Spain	493	476	491
Sweden	516	510	512
Switzerland	494	529	496
USA	504	493	499
arithmetic mean	493.45	493.16	492.61
standard deviation	33.31	46.83	37.67

We want to give a summary of all skills and all 31 countries in a scatter

188

plot. We extract exactly two principal components(PC). The linear combinations of the standardized variables ZReading, ZMath,ZScience are:

$$1.\text{PC} = 1.508*\text{ZReading} - 1.558*\text{ZMath} + 0.790*\text{ZScience}$$
$$2.\text{PC} = -1.155*\text{ZReading} + 2.221*\text{ZMath} - 0.360*\text{ZScience}$$

Countries with high points in reading as well as in natural science and few points in math have high scores of the first principal component. Countries with many points in math and few points in both reading and natural science have high scores of the second principal component.

We take the first principal component as the x-axis and the second principal component as the y-axis:

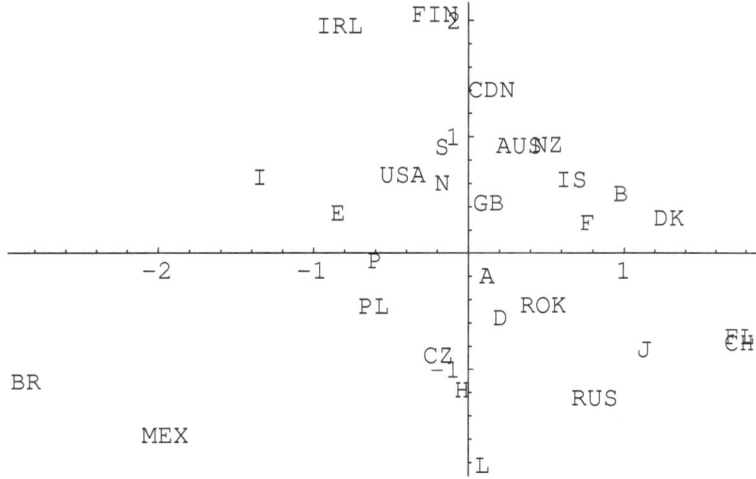

The countries (Canada, Finland) lying right above in the scatter plot come off best. The countries (Brazil, Mexico) left down performed worst.

Conclusion: We will extract only one principal component if we want to order all cases. And we will extract only two principal components if the data set is to be displayed in a scatter plot.

Finally, we have to decide whether ordering along the first principal component or the scatterplot with the first two principal components reflect well the original data set. There are three criteria for this:

1. Criterion of Kaiser (1960): In the output file of SPSS are the shares of principal components of the total variance. The extracted principal components should represent more than 90% of the total variance. In the example 12.2 about 95.573% of the total variance is represented of the first principal component and about 98.504% of the total variance is represented of the first two principal components. Therefore, there is only a little loss of information both when ordering the countries along the first principal component and displaying the countries in a two dimensional scatter plot with the first two principal components as axes.

2. Criterion of Cattell (1966): At least so many principal components are to extract as the amount of variables lying in the scree plot left side the big kink. In the scree plot of the example 12.2 there is only one variable left side of the big kink. Therefore it is sufficient (i.e. the loss of information is small) to extract just one principal component.

3. Criterion: At least so many principal components are to extract as there are eigenvalues larger than one. In the example 12.2 the values of the three eigenvalues are: 2.867, 0.088 and 0.045. There is only one eigenvalue larger than one. Therefore it is sufficient (i.e. the loss of information is small) to extract just one principal component.

We will extract only either one (for ordering) or two (for a scatter plot) principal components).

We want to know which country has the best literacy overall three variables, which country comes next and so on.

Example 12.3
The results of the principal components analysis for the example 12.2 of the PISA survey 2000 indicate only one eigenvalue larger than 1. This means that the first principal components reflects all three literacy values.

If we extract only one principal components we get the following values b_1, b_2, b_3 of the linear combination:

$$
\begin{aligned}
b_1 &= 0.341 \quad \text{Reading} \\
b_2 &= 0.339 \quad \text{Mathematics} \\
b_3 &= 0.343 \quad \text{Natural Science}
\end{aligned}
$$

The order of the 31 countries is based on the linear combination of the standardized values. We get for example for Japan the following value of the first principal component:

$$
(0.341) \cdot \frac{522-493.45}{33.31} + 0.339 \cdot \frac{557-493.16}{46.83} + 0.343 \cdot \frac{550-492.61}{37.67}
$$
$$
= 1.276964 \approx 1.28
$$

The principal components analysis reveals the following order of the 31 countries:

Country	1. Comp.	Country	1. Comp.	Country	1. Comp.
Japan	1.28	France	0.36	Poland	-0.40
Finland	1.26	Iceland	0.32	Italy	-0.46
South Korea	1.25	Switzerland	0.30	Russia	-0.73
Canada	1.03	Norway	0.23	Greece	-0.82
New Zealand	1.00	Czech Republic	0.19	Portugal	-0.83
Australia	0.96	USA	0.17	Latvia	-0.88
Great Britain	0.92	Denmark	0.08	Luxembourg	-1.33
Ireland	0.60	Liechtenstein	-0.11	Mexico	-2.14
Austria	0.54	Spain	-0.14	Brazil	-3.22
Sweden	0.53	Hungary	-0.14		
Belgium	0.36	Germany	-0.17		

Japan, Finland, South Korea, Canada performed very well, Luxembourg, Mexico and Brazil do poorly.

12.4 Summary

The principal components analysis enables us to give a summary of a data set in a two dimensional scatter plot of the first two principal components as axes. Further the principal component analysis provides an order of the cases based on the scores of the first principal component.

We extract only one principal component to order the cases of a data set. We extract exactly two principal components to construct a scatter plot for a display of all cases.

12.5 SPSS-Commands

12.5.1 Principal Components Analysis

1) Please open the file "Pisa-Studie-Hauptkomp.sav"

2) Analyze→ Dimension Reduction → Factor

3) Variables = "Reading"
"Mathematics"
"Science"

4) Select "Descriptives".
Select "'Coefficients". Click "Continue".

5) Select "Extraction".
As "Method" in the Dropdown-Menu select "Principal components".
Select "Correlation matrix".
Select "Fixed number of factors" and select "Factors to extract = 1",
when all cases should be are ordered along the first principal compo-
nent. Or choose the number of factors to be extracted is "2", when
all cases should be displayed in a scatter plot. (Alternatively select
"Based on Eigenvalues" and "Eigenvalues larger than 1".)
Select "Scree plot".
Click "Continue".

6) Select "Rotation".
Select "Varimax".
Select "Loading plot(s)".
Click "Continue".

7) Select "Scores".
Select "Save as variables" and "Display factor score coefficient ma-
trix".
Click "Continue".

8) Click "OK".

The scores of the first principal component are listed in the column FAC1_1. And the scores of the second principal component are listed in the column FAC2_1 in the Data View.

Output
We have extracted two principal components.

The first principal component explains 95.573% of the total variance. The first two principal components explain 98.504% of the total variance. And there is only one eigenvalue larger than 1 (id est 2.867):

Total Variance Explained

Component	Initial Eigenvalues		
	Total	% of Variance	Cumulative
1	2.867	95.573	95.573
2	0.088	2.930	98.504
3	0.045	1.496	100.000

Extraction Method: Principal Components Analysis.

There is only one principal component left side the big kink in the scree plot:

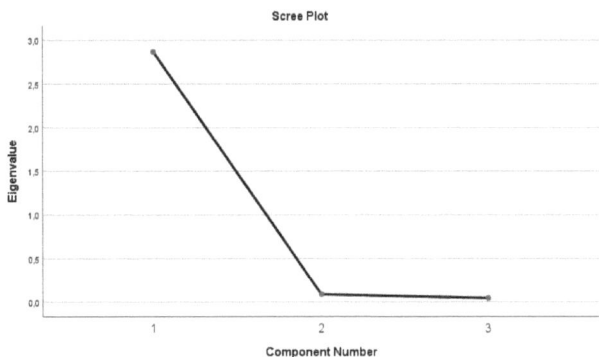

The correlations of the two extracted principal components and the three variables are listed in a table as follows:

Rotated Component Matrix

	Component	
	1	2
Reading	0.835	0.536
Mathematics	0.560	0.828
Science	0.776	0.609

The coefficients of the linear combination of the linear combination of a principal component analysis with two extracted components are listed in the table:

Component Score Coefficient Matrix

	Component	
	1	2
Reading	1.508	−1.155
Mathematics	−1.558	2.221
Science	0.790	−0.360

Extraction Method: Principal Component Analysis.
Rotation Method: Varimax with Kaiser Normalization.
Component Scores.

12.5.2 Order Based on the First Principal Component

There are different options to sort the countries in the Data View in descending order according to performance:

1. Option: In the Data View in the column select "FAC1_1" and right mouse click on "Sort Descending". Now the countries are sorted in a descending order.

2. Option:

   1) Data → Sort Cases …

2) Sort by: = FAC1_1
   Sort Order = Descending

3) "OK"

12.5.3   Scatter Plot with the First Two Principal Components

There are two options to construct a scatter plot with the first two principal components as axes:

1. Option: Select the Data View of SPSS. The values of the first principal component are listed in the column variable FAC1. The values of the second principal component are listed in the column variable FAC2. Switch over to Variable View. Rename "FAC1" as "PC1" and "FAC2" as "PC2". And rename the "Label" of FAC1 and FAC2 as "PC1" and "PC2" respectively.

1) Graphs → Legacy Dialogs → Scatter/Dot

2) Select "Simple Scatter" and click "Define".

3) Select "PC2" as *Y* Axis and "PC1" as *X* Axis.
   Label Cases by = "Land".

4) Click "Options" and select "'Display chart with case labels". Select "Continue".

5) "OK"

The countries right above in the scatter plot are countries with the highest points in literacy.

2. Option:

1) Graphs → Chart Builder → ok

2) Select "Scatter/Dot".

3) Drag the picture of the "Scatter/Dot" onto the canvas.

4) Drag the variable "PC1" as X-Axis and the variable "PC2" as Y-Axis in the chart preview.

5) Groups/Point-ID $\rightarrow$ Point-ID label

6) Drag the variable "Land" into the array of the Point Label Variable.

7) ok

13 Cluster Analysis

Purpose: Attempt of grouping objects of a data set based on selected variables in such a way that objects in the same group are more similar to each other than to those in other groups. The groups are called **cluster**.

The meaning of "cluster" is the astronomic notation of an accumulation of stars in the sky. In a scatter plot of the first and second principal component the points are gathering in groups like stars in the sky.

The cases in one and the same cluster are considered homogeneous, the groups among themselves as heterogeneous.

Remark: In contrast to the cluster analysis, the groups are already known at a discriminant analysis. A discriminant analysis assigns a new case/object to one of the existing groups.

13.1 Hierarchical Cluster Analysis

The basic idea of a cluster analysis is to compute the distance between all objects of a data set. Those objects are put together in a cluster that are the closest to each other. The levels of the variables considered in a hierarchical cluster analysis are as follows:

> nominal: yes, but only binary
> ordinal: yes   ).
> scale: yes

The variables in a hierarchical cluster analysis can be leveled binary. More precisely: If there is one binary leveled variable or more, all other variables must be binary leveled too. Mixing binary and scale leveld variables is not possible (c.f. Brosius [2018]).

Example 13.1 (univariate data set, Source: Handl p. 363)
The age of $n = 6$ persons $P_1, P_2, P_3, P_4, P_5, P_6$ is:

43 38 6 47 37 9

If we plot the six values on the axis, we get:

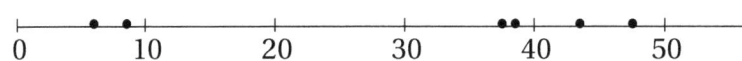

We identify two groups: the kids $\{P_3, P_6\}$ and the adults $\{P_1, P_2, P_4, P_5\}$.

Instead of tasking visually the groups we need a procedure to get the groups analytically. We compute the differences in age of the six persons P_1, P_2, P_3, P_4, P_5, P_6 with the Euclidean distance. The values are listed in the so called **proximity matrix**:

	P_1	P_2	P_3	P_4	P_5	P_6
P_1	0	5	37	4	6	34
P_2	5	0	32	9	1	29
P_3	37	32	0	41	31	3
P_4	4	9	41	0	10	38
P_5	6	1	31	10	0	28
P_6	34	29	3	38	28	0

To find the number of clusters the algorithm starts with $n = 6$ clusters, one cluster for every object:

Start of the algorithm : $\{P_1\}, \{P_2\}, \{P_3\}, \{P_4\}, \{P_5\}, \{P_6\}$

Step by step two clusters are connected until there is only one cluster (agglomeration schedule).

The distance between two clusters $\{P_1\}$ and $\{P_2\}$ is the difference in age. But how to compute the distance between the two clusters $\{P_1\}$ and $\{P_2, P_5\}$? There different methods to solve this problem. We take the single linkage solution.

First we calculate the difference in age between P_1 and P_2, that are five years, and the difference in age between P_1 and P_5, that are six years. As a distance between the clusters $\{P_1\}$ and $\{P_2, P_5\}$ we take the minimum $\min\{5, 6\} = 5$ years.

First stage of the algorithm: We join the clusters $\{P_2\}$ and $\{P_5\}$ into one cluster $\{P_2, P_5\}$, because the smallest distance in age between all six persons is the distance in age between P_2 and P_5:

    1. Decomposition: $\{P_1\}, \{P_3\}, \{P_4\}, \{P_6\}, \{P_2, P_5\}$

Second stage of the algorithm: We join the two clusters with the smallest distance. The distances are:

       $\{P_1\}$ and $\{P_3\}$ with the distance 37
       $\{P_1\}$ and $\{P_4\}$ with the distance 4
       $\{P_1\}$ and $\{P_6\}$ with the distance 34
       $\{P_3\}$ and $\{P_4\}$ with the distance 41
       $\{P_3\}$ and $\{P_6\}$ with the distance 3
       $\{P_4\}$ and $\{P_6\}$ with the distance 38
       $\{P_1\}$ and $\{P_2, P_5\}$ with the distance 5 (see above)
       $\{P_3\}$ and $\{P_2, P_5\}$ with the distance $\min\{32, 31\} = 31$
       $\{P_4\}$ and $\{P_2, P_5\}$ with the distance $\min\{9, 10\} = 9$
       $\{P_6\}$ and $\{P_2, P_5\}$ with the distance $\min\{29, 28\} = 28$

We join $\{P_3\}$ and $\{P_6\}$ with the smallest distance 3:

2. Decomposition: $\{P_1\}, \{P_4\}, \{P_3, P_6\}, \{P_2, P_5\}$

Third stage of the algorithm: We join the two clusters with the smallest distance. The distances are:

$\{P_1\}$ and $\{P_4\}$ with the distance 4
$\{P_1\}$ and $\{P_3, P_6\}$ with the distance min$\{37, 34\} = 34$
$\{P_1\}$ and $\{P_2, P_5\}$ with the distance 5 (see stage 2)
$\{P_4\}$ and $\{P_3, P_6\}$ with the distance min$\{41, 38\} = 38$
$\{P_4\}$ and $\{P_2, P_5\}$ with the distance 9 (see stage 2)
$\{P_3, P_6\}$ and $\{P_2, P_5\}$ with the distance min$\{31, 28\} = 28$

The value 31 is the distance between $\{P_3\}$ and $\{P_2, P_5\}$ and the value 28 is the distance between $\{P_6\}$ and $\{P_2, P_5\}$.

We join $\{P_1\}$ and $\{P_4\}$ with the smallest distance 4:

3. Decomposition: $\{P_1, P_4\}, \{P_3, P_6\}, \{P_2, P_5\}$

Forth stage of the algorithm: We have to join the two clusters with the smallest distance. The distances are:

$\{P_1, P_4\}$ and $\{P_3, P_6\}$ with the distance min$\{34, 38\} = 34$
$\{P_1, P_4\}$ and $\{P_2, P_5\}$ with the distance min$\{5, 9\} = 5$
$\{P_3, P_6\}$ and $\{P_2, P_5\}$ with the distance 28 (see stage 3)

We have to join $\{P_1, P_4\}$ and $\{P_2, P_5\}$ due to their smallest distance 5:

4. Decomposition: $\{P_1, P_2, P_4, P_5\}, \{P_3, P_6\}$

Stage 5 of the algorithm (last stage): We have to join the two clusters of decomposition 4. The minimal distance is min$\{34, 28\} = 28$; the value 34 is the minimal distance between $\{P_1, P_4\}$ and $\{P_3, P_6\}$ (see stage 4) and the value 28 is the minimal distance between $\{P_3, P_6\}$ and $\{P_2, P_5\}$ (see stage 4).

To decide what decomposition is the correct decomposition we look at the distances of the merging:

Person	Decomposition	min. distance
P_1, P_2	4.	5
P_1, P_3	5.	28
P_1, P_4	3.	4
P_1, P_5	4.	5
P_1, P_6	5.	28
P_2, P_3	5.	28
P_2, P_4	4.	5
P_2, P_5	1.	1
P_2, P_6	5.	28
P_3, P_4	5.	28
P_3, P_5	5.	28
P_3, P_6	2.	3
P_4, P_5	4.	5
P_4, P_6	5.	28
P_5, P_6	5.	28

These distances $\{1, 3, 4, 5, 28\}$ of the first occurrence of the objects P_i, P_j in one cluster are listed in the so called **cophenetic matrix**:

	P_1	P_2	P_3	P_4	P_5	P_6
P_1	0	5	28	4	5	28
P_2	5	0	28	5	1	28
P_3	28	28	0	28	28	3
P_4	4	5	28	0	5	28
P_5	5	1	28	5	0	28
P_6	28	28	3	28	28	0

A graphical summary of the distances $\{1, 3, 4, 5, 28\}$ of the first occurrence of two objects P_i, P_j in one cluster is the so called **dendrogram**:

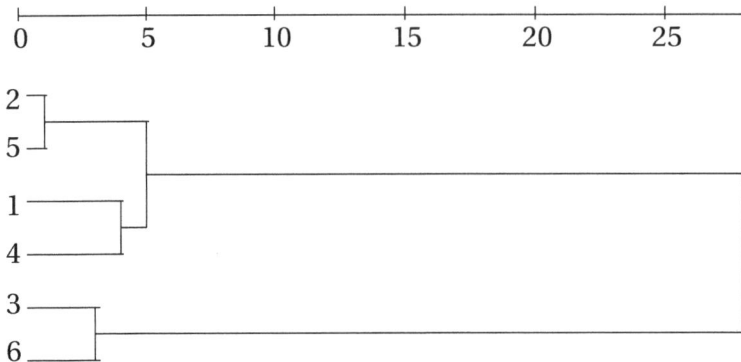

The objects are listed along the left vertical axis. The horizontal axis shows the distance between clusters when they are joined.

To determine the number of clusters in a dendrogram is a subjective process. Generally, you begin by looking for "gaps" between joinings along the horizontal axis. The biggest jump is between distance 5 and distance 28. If this joining is missing, the objects are split into two clusters:

$$\{P_1, P_2, P_4, P_5\}, \{P_3, P_6\}$$

An other method to determine the number of clusters is the biggest jump of the coefficients in the table "Agglomeration Schedule". The biggest jump of the coefficients is in stage 4 to stage 5: from 5 to 28. The number of clusters is: sample size minus number of the stage previous to the biggest jump = $6 - 4 = 2$ cluster.

Remark: The SPSS output shows a little bit different dendrogram. The distances are rescaled. In example 13.1 the distance 5 is rescaled to about $5 \cdot \frac{25}{28} = 4.46$. And the distance 28 is rescaled to about $28 \cdot \frac{25}{28} = 25$.

We consider this procedure for a multivariate data set.

Example 13.2 (multivariate data set)
PISA is a three-year-cycle survey of the knowledge and skills of 15-year-olds in OECD and other participating countries. An international comparison of

educational systems is sought in order to adapt successful forms of learning if necessary. The well-being of the learner should always be kept in mind. Each cycle comprises the three domains of Reading, Mathematical and Science literacy. In the year 2000 about 31 countries participate on the first PISA survey. The observed variables are listed in example 12.2.

There are three observed values for every country. We attempt to classify the countries into groups/clusters. The number of the clusters is unknown. The countries in the same cluster should be similar, but the countries in different clusters should be dissimilar.

SPSS plots the diagram:

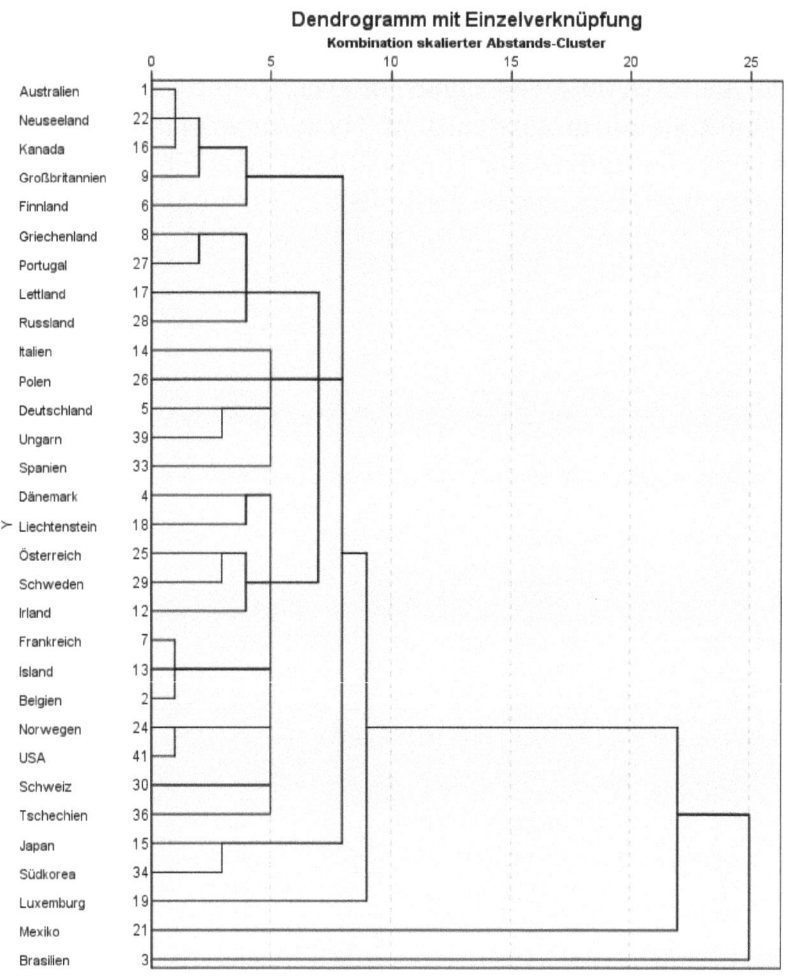

In the dendrogram of SPSS the biggest jump occurs near the country of Norway from the rescaled distance $29.445 \cdot \frac{25}{75.459} = 9.76$ to the rescaled distance $65.445 \cdot \frac{25}{75.459} = 21.68$. Therefore this line in the dendrogram is deleted:

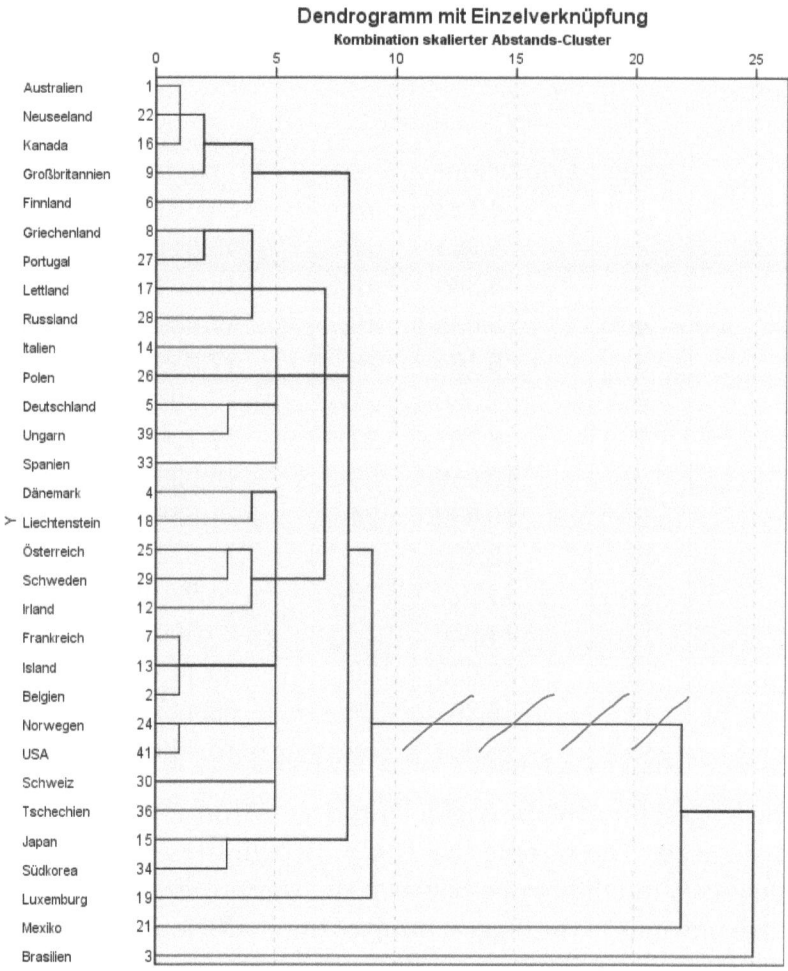

Dendrogramm mit Einzelverknüpfung
Kombination skalierter Abstands-Cluster

Australien	1
Neuseeland	22
Kanada	16
Großbritannien	9
Finnland	6
Griechenland	8
Portugal	27
Lettland	17
Russland	28
Italien	14
Polen	26
Deutschland	5
Ungarn	39
Spanien	33
Dänemark	4
Liechtenstein	18
Österreich	25
Schweden	29
Irland	12
Frankreich	7
Island	13
Belgien	2
Norwegen	24
USA	41
Schweiz	30
Tschechien	36
Japan	15
Südkorea	34
Luxemburg	19
Mexiko	21
Brasilien	3

We receive the three clusters:

1. cluster: all countries except Mexico and Brazil
2. cluster: Mexico
3. cluster: Brazil

Another method to get the number of clusters is the table with the coefficients:

Stage	Coefficient	Stage	Coefficient
1	4.123	16	15.652
2	5.385	17	15.811
3	5.385	18	16.062
4	6.083	19	17.059
5	6.164	20	17.059
6	7.550	21	17.205
7	8.307	22	17.720
8	10.050	23	18.028
9	10.630	24	21.656
10	12.450	25	23.622
11	13.077	26	24.920
12	14.866	27	26.306
13	15.033	28	29.445
14	15.297	29	65.445
15	15.524	30	75.459

The biggest jump of the coefficients occurs from 29.445 in stage 28 to 65.445 in stage 29. The sample size is $n = 31$ countries/objects. The number of clusters is the difference between the sample size and the number of the stage previous to the biggest jump of the coefficients, here $n-28 = 31-28 = 3$ clusters.

To know which countries are in the same cluster, a hierarchical cluster analysis must be run again, but now with the number of clusters is three. The cluster membership is listed in the column CLU in the Data View for every case.

Which of these three clusters represents the top group, which is the midfield and who is lagging behind? That can only be seen from the fact that the average values of the three measured competences in each clusters are calculated:

Cluster	Literacy		
	Reading	Math	Science
1	499.28	502.31	499.10
2	396	334	375
3	422	387	422

Thus, the first cluster represents the top group, the second cluster with Brazil the tail light, and the third cluster with Mexico the midfield.

If the sample are observed values only from dichotomous variables we can run the hierarchical cluster analysis, too.

Example 13.3 (Quelle: Arrenberg, Kowalski [2007])
The following question was asked in a poll about the different types of learning: Learning is easy for me if ... *(two answers only)*

☐ ... I can see the contents. (eye-minded type of learning)

☐ ... I can discuss the results with other students. (communicative type of learning)

☐ ... I can give a trial. (primary activity type of learning)

☐ ... I can explain the contents to somebody. (auditive type of learning)

The four dichotomous variables (eye-minded, communicative, primary activity, auditive) are decoded as 0=no and yes=1, especially the four variables are binary variables.

Abbreviation for the four types of learning: a=auditive, k=communicative, m=primary activity, v=eye-minded. We get the following frequencies of 933 respondents:

Type of Learning	a	22
	ak	96
	am	59
	av	195
	k	35
	m	23
	mk	91
	v	94
	vk	175
	vm	143
Total		933

To run a hierarchical cluster analysis with SPSS we select the following method: "Within groups linkage". And as a measure we select "Binary" with the difference "Pattern difference". The biggest jump in the table of the coefficients occurs from stage 931 to stage 932:

Stage	Coefficient
931	0.064
932	0.113

The number of clusters are the difference between the sample size $n = 933$ and the number of the stage previous to the biggest jump: $933 - 931 = 2$ clusters. Though we discriminate between two types of learning:

Type of learning ∗ Cluster Cross table

Total

		Cluster		Frequency
		1	2	
Type of learning	a	6	16	22
	ak	0	96	96
	am	0	59	59
	av	195	0	195
	k	10	25	35
	m	10	13	23
	mk	0	91	91
	v	94	0	94
	vk	175	0	175
	vm	143	0	143
Total		633	300	933

We comment the two clusters due to the cross table as:

Cluster 1: Type of learning av, vk, vm, v
Cluster 2: Type of learning ak, mk, am

13.2   K-Means Cluster Analysis

In a hierarchical cluster analysis the number of clusters is a result of the analysis.

In contrast the number of clusters must be known in advance to run a k-means cluster analysis. The variables of the k-means cluster analysis must be leveled as follows:

nominal: no
ordinal: no
scale: yes

Example 13.4 (univariate data set, c.f. Handl p. 387)
The age was observed for $n = 6$ persons P_1, P_2, P_3, P_4, P_5, P_6: 43, 38, 6, 47, 37, 9 years.

The six persons/cases should be classify into $k = 2$ clusters by the k-means cluster analysis.

Procedure: The k-means cluster analysis starts with an arbitrary decomposition into two classes:

Initial decomposition $\{P_1, P_2, P_3\}, \{P_4, P_5, P_6\}$

The average age of the first class is $\bar{x} = \frac{1}{3}[43 + 38 + 6] = 29$ years and the average age of the second class is $\bar{x} = \frac{1}{3}[47 + 37 + 9] = 31$ years. As a measure of the goodness of the decomposition we take the total of the squared distances (Euclidean distance) of the objects in a class from the average values of the class:

$$
\begin{array}{rcl}
(43 - 29)^2 & = & 196 \\
(38 - 29)^2 & = & 81 \\
(6 - 29)^2 & = & 529 \\
(47 - 31)^2 & = & 256 \\
(37 - 31)^2 & = & 36 \\
(9 - 31)^2 & = & 484 \\
\hline
\Sigma & & 1\,582
\end{array}
$$

We get the total $1\,582$. Now the objects of the two classes were changed to get a smaller total than $1\,582$. First we put P_1 into the second class:

1. Decomposition: $\underbrace{\{P_2, P_3\}}_{\bar{x}=22}, \underbrace{\{P_1, P_4, P_5, P_6\}}_{\bar{x}=34}$

The squared Euclidean distances are:

210

$$
\begin{array}{rcl}
(38-22)^2 & = & 256 \\
(6-22)^2 & = & 256 \\
(43-34)^2 & = & 81 \\
(47-34)^2 & = & 169 \\
(37-34)^2 & = & 9 \\
(9-34)^2 & = & 625 \\
\hline
\Sigma & & 1\,396
\end{array}
$$

We get a smaller total of 1 396. Now we continue changing the objects until the total is the minimal total:

Last decomposition: $\underbrace{\{P_3, P_6\}}_{\bar{x}=7.5}, \underbrace{\{P_1, P_2, P_4, P_5\}}_{\bar{x}=41.25}$

The squared Euclidean distances are:
$$
\begin{array}{rcl}
(6-7.5)^2 & = & 2.25 \\
(9-7.5)^2 & = & 2.25 \\
(43-41.25)^2 & = & 3.0625 \\
(38-41.25)^2 & = & 10.5625 \\
(47-41.25)^2 & = & 33.0625 \\
(37-41.25)^2 & = & 18.0625 \\
\hline
\Sigma & & 69.25
\end{array}
$$

The total of 69.25 is the minimal total. Though we have found the two clusters. One cluster is the cluster with the two kids and the other cluster is the cluster with the four adults. This split into two clusters is identical to the two clusters of hierarchical clustering example 13.1.

In particular, for larger sample sizes than $n = 6$, the cluster memberships of the hierarchical and k-means cluster analysis often are different.

The decision of SPSS as to when the sum of the squared differences is minimal is made by skillful trial and error. Esp. otherwise, if the number of iterations is too small, we get different results. We all get the same cluster splits, if the number of iterations is increased from 10 to 35.

If it is not clear how many clusters should be constructed, a hierarchical cluster analysis should be carried out beforehand. The hierarchical cluster analysis reveals the reasonable number of clusters.

Example 13.5 (multivariate data set, c.f. example 13.2)
PISA is a three-year-cycle survey of the knowledge and skills of 15-year-olds
in OECD and other participating countries. Each cycle comprises the three
domains of Reading, Mathematical and Science literacy. In the year 2000
about 31 countries participate on this survey. The variables are:

- Reading literacy (measured in points)

- Mathematical literacy (measured in points)

- Science literacy (measured in points)

For every county/object we have three observed values. We want to classify the 31 countries into three groups. The clusters of the k-means cluster
analysis are:

Country	Cluster	Country	Cluster	Country	Cluster
Australia	1	Italy	2	Poland	2
Belgium	1	Japan	1	Portugal	2
Brazil	3	Canada	1	Russia	2
Denmark	1	South-Korea	1	Sweden	1
Germany	2	Latvia	2	Switzerland	1
Finland	1	Liechtenstein	2	Spain	2
France	1	Luxembourg	2	Czecho	1
Greece	2	Mexico	3	Hungary	2
Great Britain	1	New Zealand	1	USA	1
Ireland	1	Norway	1		
Iceland	1	Austria	1		

We consider the average values within a cluster to comment it. The average
values within cluster are called **cluster centers**:

Final Cluster Centers

	Cluster		
	1	2	3
Reading	515	474	409
Mathematics	521	471	361
Science	516	472	399

All arithmetic means in the first cluster have the largest values. Therefore the 18 countries of the first cluster are the excellence cluster. The smallest arithmetic values are in the third cluster. Therefore the two countries of the third cluster bring up the rear. The eleven countries in the second cluster have a moderate level of literacy.

After having run a cluster analysis we can display the clusters in a scatter plot based on the first two principal components.

Example 13.6 (*Kriminalitaet.sav* c.f. Schlittgen [2009] p. 5)
The crime is continuously recorded in the 50 states of the US plus Hawaii ($n = 51$). We consider the following seven delicts:

$X_1 =$	Murder	$X_5 =$	Burglary
$X_2 =$	Rape	$X_6 =$	Thievery
$X_3 =$	Robbery	$X_7 =$	Car Theft
$X_4 =$	Criminal Assault		

So many cases of the seven delicts per 100 000 inhabitants were discovered in the year 2002:

State	X_1	X_2	X_3	X_4	X_5	X_6	X_7
Alabama	7	37	133	268	949	2762	310
Alaska	5	79	76	403	607	2755	384
Arizona	7	30	147	370	1083	3694	1057
Arkansas	5	28	93	298	857	2625	251
Californ	7	29	185	373	679	2038	633
Colorado	4	46	79	223	703	2778	514
Connecti	2	21	117	170	494	1858	334
Delaware	3	44	143	409	663	2298	379
Columbia	46	46	672	869	906	3802	1681
Florida	6	40	195	529	1061	3060	530
Georgia	7	25	157	270	864	2740	444
Hawaii	2	30	97	133	1022	3964	796
Idaho	3	37	18	197	555	2167	196
Illionoi	8	34	201	379	644	2396	356
Indiana	6	30	107	214	692	2372	329
Iowa	2	27	40	217	635	2330	198
Kansas	3	38	80	256	725	2720	266
Kentucky	5	27	75	173	681	1729	214
Louisian	13	34	159	456	1012	2974	450
Maine	1	29	21	57	538	1900	110
Maryland	9	25	246	490	729	2626	623
Massachu	3	28	112	343	517	1679	414
Michigan	7	53	118	362	706	2133	495
Minnesot	2	45	78	142	559	2433	276
Mississi	9	39	117	178	1031	2454	332
Missouri	6	26	124	383	753	2819	492
Montana	2	26	31	293	362	2604	196
Nebraska	3	27	79	206	597	2975	371
Nevada	8	43	236	351	872	2184	805
NeHampsh	1	35	32	93	379	1527	153
NeJersey	4	16	162	193	511	1723	416
NeMexico	8	55	119	557	1058	2879	401
NewYork	5	20	191	280	400	1660	247
NCarolin	7	26	147	291	1196	2756	299
NDakota	7	26	147	291	1196	2756	299
Ohio	5	42	157	148	868	2513	375
Oklahoma	5	45	85	369	1007	2868	366
Oregon	2	35	78	177	730	3377	469
Pennsylv	5	30	139	228	451	1722	266
RhodeIsl	4	37	86	159	600	2248	456
SCarolin	7	48	141	627	1065	3000	411
SDakota	1	47	15	113	399	1595	108
Tennessee	7	40	162	508	1057	2788	458
Texas	6	39	173	361	976	3163	471
Utah	2	41	49	145	653	3229	333
Vermont	2	20	13	72	566	1733	125
Virginia	5	25	95	166	435	2160	253
Washingt	3	45	96	202	905	3189	667
WVirgini	3	18	37	176	537	1528	216
Wiscons	3	23	87	113	513	2267	247
Wyoming	3	30	19	222	491	2668	149

We want to classify the states into crime clusters. Finally, these clusters should be entered in a graph to see which one of the states are concerned about crime and which states are clear on crime differ.

At first we run a hierarchical cluster analysis with all seven variables (first cluster analysis). The biggest jump of the coefficients occurs from stage 49 to stage 50:

Stage	Coefficient
49	685.474
50	979.169

Thus the recommended number of clusters is $n - 49 = 51 - 49 = 2$.

We run again a hierarchical cluster analysis (second cluster analysis) to get the cluster membership of each state, but this time with the following SPSS commands:

1) Click "Statistics ...".
   Single Solution.
   "Number of Clusters = 2".
   Click "Continue".

2) Click "Save ...".
   Single Solution.
   Select "Number of Clusters = 2", to get the cluster membership of each state.
   Click "Continue".

It turns out that there is a cluster only with the District of Columbia and a cluster with all the other 50 states. In the District of Columbia, the crime is so high that there is no comparable state given. However, this does not discriminate between the 50 states, so we remove the District of Columbia from the data setand run again a hierarchical cluster analysis (third cluster analysis).

The biggest jump of the coefficients occurs from stage 48 to stage 49:

Stage	Coefficient
48	451.039
49	685.474

Thus the number of clusters is $n - 48 = 50 - 48 = 2$.

To see which of the two clusters is lower in crime, we run a k-means cluster analysis (fourth cluster analysis). The given number of clusters is 2. The average values (cluster centers of the final solution) of the seven delicts in the two clusters are:

Delikc	Cluster 1	2
Murder	4	6
Rape	31	38
Robbery	97	121
Criminal Assault	220	322
Burglary	558	892
Thievery	2012	2928
Car Theft	309	443

From the average values, it is clear that the first cluster (California, Connecticut, ..., Wisconsin) are low-crime states and the second cluster (Arizona, Hawaii, ..., Wyoming) are states with high levels of crime.

To plot the clusters in a scatter diagram, we run a principal component analysis with all seven variables and extract exactly two principal components. The first two principal components account for nearly 70% (exactly 69.374%) of the total variance, that is not a lot, that is, the loss of information through considering only the two major components instead of the seven variables, is about 30%. Nevertheless we will draw the two clusters by hand:

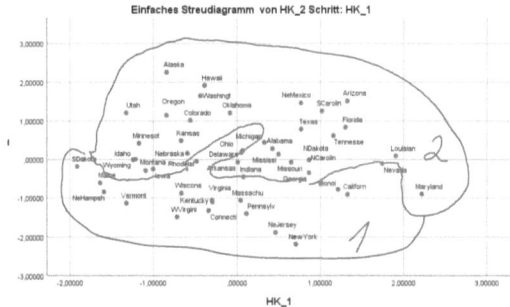

In the scatter plot the low-crime states are lying bottom left and the high-crime states are lying top right.

13.3 Two-Step Cluster Analysis

The Two-Step cluster analysis classifies the objects of a data set based on nominal or ordinal or scale leveled variables: :
>   nominal: yes
>   ordinal: yes
>   scale: yes

Discrete variables that are nominal or ordinal leveled are denoted as **categorial** variables. Continuous variables are scale leveled variables. Vice versa a scale leveled variable must not be a continuous variable. For example the variable X="Number of siblings" is a scale leveled discrete variable.

The assumptions of the Two-Step cluster analysis are:

1) Stochastic independence of all variables
   The Chi-square test (confer chapter 3.1) checks the stochastic independence of two variables; if necessary, the values of a scale leveled variable must be recoded into classes so that the rule of thumb is fulfilled.
   But the pairwise stochastic independence doesn't include the stochastic independence of all variables. This means that there is no procedure in SPSS to check the assumption of independence. Therefore

217

we must assume that all variables are stochastically independent.

2) Normal distribution of every scale leveled variable
The goodness of fit test of Shapiro-Wilk or Lilliefors verifies this assumption. (confer chapters 5.1 and 5.2)

3) Multinomial distribution for every not scale leveled variable
A random variable $X_1,\ldots,X_k$ has a multinomial distribution if there are $k+1$ distinct possible outcomes of a trial and let $p_1,p_2,\ldots,p_{k+1}$ denote these probabilities for theses outcomes. Suppose that we repeat the trial n times. If the trials are repeated and stochastically independent and if the probabilities $p_1,p_2,\ldots,p_{k+1}$ are unchanged in every trial. (In case of only two distinct outcomes, i.e. $k=1$, the multinomial distribution is identic with the binomial distribution.) How to check this assumption with SPSS? If the probabilities p_1, p_2, $\ldots$, p_{k+1} are known, a Chi-Square test of goodness-of-fit verifies the null hypothesis of multinomial distribution (Nonparametric Tests $\rightarrow$ One Sample $\rightarrow$ Automatically compare observed data to hypothesized). The p-value is computed based on a Chi-Square distribution with k degrees of freedom. But in the k-means cluster analysis the probabilities $p_1,p_2,\ldots,p_{k+1}$ are unknown and $m=k$ of these probabilities must be estimated. In this case of unknown estimated probabilities the p-value of the Chi-Square Goodness-of-fit test has to be computed based on the chi-square distribution with $k-m=0$ degree of freedom and such a distribution is not defined. Therefore SPSS provides no procedure to check the assumption of multinomial distribution, so we have to suppose multinomial distribution of every nominal or ordinal leveled variable.

The number of clusters is optional for the Two-Step cluster analysis. Either the number is known in advance or the number of clusters is a result of the analysis.

Example 13.7 (*EU Staaten_2013.sav*)
The member states of the EU in the year 2013 should be classified based on their inflation rate, deficit of the overall budget (in bn Euro), the national debt/GDP (gross domestic product) ratio and participation (yes/no) for the

Eurozone in the year 2013:

Member state	Inflation	Deficit	debt	Eurozone
Germany	1.90	2150.50	81.70	yes
Italy	1.40	2034.76	126.10	yes
France	1.00	1870.29	89.90	yes
GB	2.70	1638.72	88.70	no
Spain	2.20	922.83	85.30	yes
Netherlands	3.20	431.36	68.70	yes
Belgium	1.50	394.22	99.60	yes
Greece	−0.30	305.29	161.30	yes
Austria	2.20	231.59	74.60	yes
Poland	0.20	219.54	53.80	no
Portugal	1.20	208.28	119.70	yes
Ireland	0.70	204.05	118.00	yes
Sweden	.50	168.72	38.60	no
Denmark	.60	109.67	45.30	no
Finland	2.30	105.27	53.50	yes
Hungary	2.00	76.67	78.60	no
Czech Republic	1.60	70.84	43.90	no
Rumania	4.50	52.04	37.20	no
Slovakia	1.70	39.35	48,.60	yes
Slovenia	2.20	19.12	53.20	yes
Cyprus	.80	15.34	80.90	yes
Lithuania	1.30	13.53	38.50	no
Luxembourg	2.00	10.04	18.40	yes
Latvia	.20	8.76	44.00	no
Bulgaria	1.20	7.21	17.90	no
Malta	.60	5.17	77.00	yes
Estonia	4.10	1.72	8.00	yes
Croatia	2.20	4.00	52.10	no

The Two-Step cluster analysis classifies two clusters:

Cluster 1: Members of the Eurozone
Cluster 2: Members outside of the Eurozone

The measure of cohesion quantifies the average distance of two points in the same cluster. The measure of separation quantifies the minimal distance of two clusters. The value of the silhouette measure is lying in the interval $[-1; +1]$. The silhouette measure is a combination of the measure of cohesion and the measure of separation, values close to one indicates a good classification. Here the value of the measure of silhouette is close to 0.5, that indicates a fair classification.

Remark: If the scheduling of the cases is changed we will get different results of the cluster analysis. Therefore it is recommended to order the cases by random before running a Two-Step cluster analysis. A changing of the order of the cases will not effect the results of the cluster analysis if the sample size is large, rule of thumb: $n \geq 200$.

13.4 Summary

The three cluster analyses (hierarchical, k-means, two-step) classifies all cases in a few clusters.

Hierarchical clustering can be used to determine the number of clusters into that a data set is to be decomposed. For the subsequent interpretation of the clusters, the k-means cluster analysis is more convenient than the hierarchical cluster analysis, because the k-means cluster analysis calculates the average values of the variables considered in each cluster. Found clusters can be represented visually in a scatter plot with the first two principal components as axes.

The two-step cluster analysis requires a lot, which unfortunately cannot be verified with SPSS.

13.5 SPSS-Commands

13.5.1 Hierarchical Cluster Analysis

1) Please open the file "Pisa-Studie-2000.sav"

2) Analyze → Classify → Hierarchical Cluster ...

3) Select "Reading", "Mathematics" and "Science" as analysis variables.

4) Select "Plots".
   Choose "Dendrogram". Select "None" in the Icicle group. Select "Continue".

5) Select "Method".
   Select "Nearest neighbor" as the cluster method in the dropdown-menu. Select "Interval" and select "Euclidean distance" as a measure of distance in the dropdown-menu. In the dropdown-menu "Transform Values" select "None". Click "Continue".
   (If there are ordinal leveled variables please select cluster method as "Furthest neighbor" and measure as "Minkowski" with "$q=1$". If there are binary variables please select cluster method as "Within groups linkage" and measure as "binary" and "Pattern difference".)

6) Select "Statistics".
   If the number of clusters is known, select "Single solution" with "Number of clusters = ... ". Select "Continue".

7) Select "Save".
   If you want to obtain the list with the cluster membership of each object, the number of clusters must be known. Select "Single solution" with "Number of clusters = ... ". Select "Continue".
   The cluster membership is listed in the Data View in the column CLU.

8) Click "Ok" to obtain the cluster analysis.

In the output of SPSS the values of the cophenetic matrix are listed in the table "Coefficients". The biggest jump of the coefficients occurs in stage 28 from 29.445 to 65.445 in stage 29. The sample size is $n = 31$ countries. The number of clusters is $n - 28 = 31 - 28 = 3$ clusters.

The decomposition of stage 28 is the appropriate decomposition and can be seen in the dendrogram.

⚠ The distances in the dendrogram of SPSS are always transformed/rescaled distances, the largest SPSS-distance is always 25. This is a linear transformation of the real distances. You can see the biggest jump of the rescaled coefficients in the dendrogram as well.

Now, we know the appropriate number of clusters. For large sample sizes it is difficult to recognize the cluster membership of each object in the dendrogram. So we run the hierarchical cluster analysis one more time, but this time with the known number of clusters. We select "Single solution" and "Number of clusters=3" in the link "Save" and in the link "Statistics". And the cluster membership of each country is listed in the column "CLU" in the Data View.

13.5.2 K-Means Cluster Analysis

1) Please open the file "Pisa-Studie-2000.sav"

2) Analyze → Classify → K-Means Cluster ...

3) Variables = "Lesekompetenz", "MathGrundbildung" and "NaturwGrundbildung".
   Number of Clusters = 3.
   Select "Iterate and classify".

4) Click "Iterate".
   Maximum Iterations = 35.
   Click "Continue".

5) Select "Options".
   Select "ANOVA table", "Cluster information for each case" and "Ex-

clude cases listwise".
Click "Continue".

6) Click "Save".
   Select "Cluster membership".
   (If you want to know the distance of a country from the cluster center
   select "Distance from cluster center". Then the distance is listed in
   the column QCL_2. If the distance is small, the literacy of the country
   is close to the arithmetic mean in this cluster. If the distance is large,
   the literacy of the country is far away of the arithmetic mean in this
   cluster.)

7) Click "'Ok"' to obtain the findings of the k-means cluster analysis.

The arithmetic average values of the three considered variables Reading
Literacy, Mathematical Literacy, Nature Science Literacy are listed in the
table "Final Cluster Centers" separately for each cluster. The cluster membership of each case is listed in the Data View in the column QCL_1.

If there are missing values in the data set, SPSS offers two different options:

- If there is an object/case with one or more missing value, the k-means
  cluster analysis excludes this object/case from analysis. For this solution please select "Options" and "Missing values" and select "Exclude
  cases listwise".

- If there is an object/case with one or more missing value, the k-means
  cluster analysis includes all other values of this object/case. For this
  solution please select "Options" and "Missing values" and select "Exclude cases pairwise".

13.5.3 Two-Step Cluster Analysis

1) Please open the file "EU_Staaten_2013.sav"

2) Analyze → Classify → Two-Step Cluster

3) Categorical Variables = Eurozone
   Continuous Variables = Inflation, Schulden, Schulden_BIP.
   Distance Measure = Log-likelihood
   Number of clusters: "Determine automatically, Maximum = 15" (or
   Specify fixed, Number = ...)

4) Output
   Under "Working Data File" select "Create cluster membership vari-
   able" (TSC-column in editor)
   Continue

5) Click "Paste"
   Please paste in the syntax previous to the row)SAVE:
   /PRINT COUNT SUMMARY

6) In Symbol list select "Run → All"

The cluster membership of each case is listed in the column TSC of the Data
View. The average values and empirical standard deviations of considered
variables are listed in the table "Centroids":

Centroids

		Inflation		Debt		Debt_BIP	
		Mean	Std Deviation	Mean	Std Deviation	Mean	Std Deviation
Cluster	1	1.6882	1.03494	526.4224	749.58928	80.2647	38.58320
	2	1.5455	1.28403	215.4273	477.23469	48.9636	19.68361
Combined		1.6321	1.11822	404.2457	664.26268	67.9679	35.60922

14 Summary

Finally, we give an overview of the procedures covered in this book.

14.1 Overview of Questions

First of all, the question and then the associated procedure should be determined.

Question: Is there a relationship between two variables?

Method: **Pearson's Chi-Square Test**

Example: Is there a relationship between age and viral load of a disease?

Answer: If the p-value is equal to or smaller than 0.05, age and viral load depend stochastically on one another if the sample was representative.

Question: What is the relationship in a bivariate sample?

Method: **Bravais-Pearson correlation coefficient, Spearman's rho, Kendall's tau-b, Gamma coefficient, coefficient of contingency**

Example: Do older customers buy more than younger customers?

Answer: If rho resp. tau-b resp. gamma are positive and both the age classes and the sales figures are in ascending order in the crosstab, then younger customers in this sample tend to spend less, older customers tend to spend more.

Question: Is the distribution a Normal distribution?

Method: **Jarque-Bera Test, Lilliefors Test, Shapiro-Wilk Test**

Example: Are the sales Normally distributed?

Answer: If the p-value is equal to or smaller than 0.05, the distribution differs significantly from a Normal distribution if the sample was representative.

Question: Are the theoretical variances of a variable across two or more groups different?

Method: **Levene Test**

Example: Are the risks of two bonds roughly the same?

Answer: If the p-value is equal to or smaller than 0.05, the risks of the two bonds differ significantly if the sample was representative.

Question: Are the theoretical means of a variable across two groups different?

Method: **t-Test, Welch test**

Example: Are the mean sales in the two age classes under 60 years and 60+ roughly the same?

Answer: If the p-value is equal to or smaller than 0.05, the two theoretical mean sales differ significantly if the sample was representative.

Question: Which forecasted value can be expected?

226

Method: **Regression analysis**

Example: What price is appropriate for a used car with a certain mileage and a certain year of construction?

Answer: A price is predicted.

Question: Does a variable have the same distribution in three or more groups?

Method: **Analysis of variance**

Example: Are the mean sales in the three age classes younger than 18 years, 18 up to 50 years, 50+ roughly the same?

Answer: If the p-value is equal to or smaller than 0.05, at least two of the theoretical mean sales differ significantly if the sample was representative.

Question: Does a variable have the same theoretical medians in two or more groups?

Method: **Kruskal-Wallis Test**

Example: Are the median sales in the three age classes younger than 18 years, 18 up to 50 years, 50+ roughly the same?

Answer: If the p-value is equal to or smaller than 0.05, at least two of the theoretical median sales differ significantly if the sample was representative.

Question: How can we order all cases of a bivariate or multivariate data set?

Method: **Principal components analysis**

Example: Which country performed the best across the three literacies in reading, mathematics and sciences?

Answer: We extract exactly one principal component and we order all
            countries due to their values of the first principal component.

Question: How can all cases of a multivariate data set be summarized in a
            two dimensional scatter plot?

Method: **Principal components analysis**

Example: Where are the crime clusters in a scatter plot?

Answer: We extract exactly two principal components and we construct
            a scatter plot with the first two principal components as the
            axes. The low crime cluster is at the bottom left of the scatter
            plot. The high crime cluster is at the top right of the scatter plot.

Question: How to classify all cases in two or more groups?

Method: **Hierarchical cluster analysis, k-means cluster analysis, two-
            step cluster analysis**

Example: The states of the US are divided into different crime clusters.

Answer: States like Vermont, Maine, New Hampshire, Connecticut etc.
            have low crime rates. States like New Mexico, Arizona, South
            Carolina etc. have high crime rates.

14.2   Overview of the Levels of the Variables

To choose the correct statistical method you have to consider the level of
the variables. The following summary should help you to be on the right
path:

One Univariate Sample $(x_1, x_2, \ldots, x_n)$			
Level	Parameter		Testing
	Location	Dispersion	
nominal	Modus	–	
ordinal	Median	Quartiles	
scale	Average	Standard-deviation	One Sample t-Test
			Lilliefors Test, Shapiro-Wilk Test,
			Jarque-Bera Test

Two Univariate Samples $(x_1,\ldots,x_n)$ and $(y_1,\ldots,y_m)$				
Tests				
		Y		
	nominal	dichotomous	ordinal	scale
nominal				
dichotomous				
X ordinal				
scale				Independent Samples t-Test

Two univariate data sets are entered in SPSS as follows: In the data view, the values $x_1, \ldots, x_n, y_1, \ldots, y_m$ are entered in the first column. The second column then shows whether the value belongs to the sample from X (i.e. second column=1) or from Y (i.e. second column=2): $1, \ldots, 1, 2, \ldots, 2$.

Three Or More Univariate Samples $(x_1,\ldots,x_n)$, $(y_1,\ldots,y_m)$ and $(z_1,\ldots,z_p)$					
Tests					
				Z	
		nominal	dichotomous	ordinal	scale
X nominal	Y nominal				
	dichotomous				
	ordinal				
	scale				
X dichotomous	Y nominal				
	dichotomous				
	ordinal				
	scale				
X ordinal	Y nominal				
	dichotomous				
	ordinal			Kruskal-Wallis	
	scale				
X scale	Y nominal				
	dichotomous				
	ordinal				
	scale				ANOVA, Kruskal-Wallis

The input of three univariate data sets in SPSS is as follows: In the data view, in the first column the values $x_1, \ldots, x_n, y_1 \ldots, y_m, z_1, \ldots, z_p$ are entered. The second column then shows whether the value in the row belongs to the first or second or third sample: $1, \ldots, 1, 2, \ldots, 2, 3, \ldots, 3$.

One Bivariate Sample $(x_1,y_1),(x_2,y_2),\ldots,(x_n,y_n)$				
Measures of Association				
		Y		
		nominal	dichotomous	ordinal	scale
	nominal	Coefficient of Contingency	Coefficient of Contingency	Coefficient of Contingency	Coefficient of Contingency
	dichotomous	Coefficient of Contingency	Gamma Tau-b, Rho	Gamma Tau-b, Rho	Gamma Tau-b, Rho
X	ordinal	Coefficient of Contingency	Gamma Tau-b, Rho	Gamma Tau-b, Rho	Gamma Tau-b, Rho
	scale	Coefficient of Contingency	Gamma Tau-b, Rho	Gamma Tau-b, Rho	Pearson

One Bivariate Sample $(x_1,y_1),(x_2,y_2),\ldots,(x_n,y_n)$				
Testing				
		Y		
		nominal	dichotomous	ordinal	scale
	nominal	χ^2-Test	χ^2-Test	χ^2-Test	χ^2-Test
	dichotomous	χ^2-Test	χ^2-Test	χ^2-Test	χ^2-Test
X	ordinal	χ^2-Test	χ^2-Test	χ^2-Test	χ^2-Test
	scale	χ^2-Test	χ^2-Test	χ^2-Test	χ^2-Test, Paired Samples t-Test

One multivariate data Sample $(x_1,y_1,z_1),(x_2,y_2,z_2),\ldots,(x_n,y_n,z_n)$						
Regression						
unab-hängig	ab-hängig	Z unabhängig				
		nominal	binary	ordinal	scale
X nominal		nominal	MultReg	MultReg	MultReg	
		binary	BinReg	BinReg	BinReg	BinReg
	Y	ordinal	OrdReg	OrdReg	OrdReg	
		scale				
X binary		nominal	MultReg	MultReg	MultReg	
		binary	BinReg	BinReg	BinReg	BinReg
	Y	ordinal	OrdReg	OrdReg	OrdReg	
		scale				
X ordinal		nominal	MultReg	MultReg	MultReg	
		binary	BinReg	BinReg	BinReg	BinReg
	Y	ordinal	OrdReg	OrdReg	OrdReg	
		scale				
X scale		nominal				
		binary	BinReg	BinReg	BinReg	BinReg
	Y	ordinal				
		scale				LinReg

BinReg   Binary logistic Regression for the dependent variable Y and for the two independent variables X and Z

LinReg   Multiple linear Regression for the dependent variable Y and for the two independent variables X and Z

MultReg Multinomial logistic Regression for the dependent variable Y and for the two independent variables X and Z

OrdReg Ordinal Regression for the dependent variable Y and for the two independent variables X and Z

.

A Quantils of the Normal Distribution

Example: $\Phi(u) = 0.164 \;\Rightarrow\; u = -0.9782$

$\qquad\qquad u = -0.9822 \;\Rightarrow\; \Phi(u) = 0.163$

Prob.	.000	.001	.002	.003	.004	.005	.006	.007	.008	.009
0.00		−3.0902	−2.8782	−2.7478	−2.6521	−2.5758	−2.5121	−2.4573	−2.4089	−2.3656
0.01	−2.3263	−2.2904	−2.2571	−2.2262	−2.1973	−2.1701	−2.1444	−2.1201	−2.0969	−2.0749
0.02	−2.0537	−2.0335	−2.0141	−1.9954	−1.9774	−1.9600	−1.9431	−1.9268	−1.9110	−1.8957
0.03	−1.8808	−1.8663	−1.8522	−1.8384	−1.8250	−1.8119	−1.7991	−1.7866	−1.7744	−1.7624
0.04	−1.7507	−1.7392	−1.7279	−1.7169	−1.7060	−1.6954	−1.6849	−1.6747	−1.6646	−1.6546
0.05	−1.6449	−1.6352	−1.6258	−1.6164	−1.6072	−1.5982	−1.5893	−1.5805	−1.5718	−1.5632
0.06	−1.5548	−1.5464	−1.5382	−1.5301	−1.5220	−1.5141	−1.5063	−1.4985	−1.4909	−1.4833
0.07	−1.4758	−1.4684	−1.4611	−1.4538	−1.4466	−1.4395	−1.4325	−1.4255	−1.4187	−1.4118
0.08	−1.4051	−1.3984	−1.3917	−1.3852	−1.3787	−1.3722	−1.3658	−1.3595	−1.3532	−1.3469
0.09	−1.3408	−1.3346	−1.3285	−1.3225	−1.3165	−1.3106	−1.3047	−1.2988	−1.2930	−1.2873
0.10	−1.2816	−1.2759	−1.2702	−1.2646	−1.2591	−1.2536	−1.2481	−1.2426	−1.2372	−1.2319
0.11	−1.2265	−1.2212	−1.2160	−1.2107	−1.2055	−1.2004	−1.1952	−1.1901	−1.1850	−1.1800
0.12	−1.1750	−1.1700	−1.1650	−1.1601	−1.1552	−1.1503	−1.1455	−1.1407	−1.1359	−1.1311
0.13	−1.1264	−1.1217	−1.1170	−1.1123	−1.1077	−1.1031	−1.0985	−1.0939	−1.0893	−1.0848
0.14	−1.0803	−1.0758	−1.0714	−1.0669	−1.0625	−1.0581	−1.0537	−1.0494	−1.0450	−1.0407
0.15	−1.0364	−1.0322	−1.0279	−1.0237	−1.0194	−1.0152	−1.0110	−1.0069	−1.0027	−0.9986
0.16	−0.9945	−0.9904	−0.9863	−0.9822	−0.9782	−0.9741	−0.9701	−0.9661	−0.9621	−0.9581
0.17	−0.9542	−0.9502	−0.9463	−0.9424	−0.9385	−0.9346	−0.9307	−0.9269	−0.9230	−0.9192
0.18	−0.9154	−0.9116	−0.9078	−0.9040	−0.9002	−0.8965	−0.8927	−0.8890	−0.8853	−0.8816
0.19	−0.8779	−0.8742	−0.8705	−0.8669	−0.8633	−0.8596	−0.8560	−0.8524	−0.8488	−0.8452
0.20	−0.8416	−0.8381	−0.8345	−0.8310	−0.8274	−0.8239	−0.8204	−0.8169	−0.8134	−0.8099
0.21	−0.8064	−0.8030	−0.7995	−0.7961	−0.7926	−0.7892	−0.7858	−0.7824	−0.7790	−0.7756
0.22	−0.7722	−0.7688	−0.7655	−0.7621	−0.7588	−0.7554	−0.7521	−0.7488	−0.7454	−0.7421
0.23	−0.7388	−0.7356	−0.7323	−0.7290	−0.7257	−0.7225	−0.7192	−0.7160	−0.7128	−0.7095
0.24	−0.7063	−0.7031	−0.6999	−0.6967	−0.6935	−0.6903	−0.6871	−0.6840	−0.6808	−0.6776
0.25	−0.6745	−0.6713	−0.6682	−0.6651	−0.6620	−0.6588	−0.6557	−0.6526	−0.6495	−0.6464
0.26	−0.6433	−0.6403	−0.6372	−0.6341	−0.6311	−0.6280	−0.6250	−0.6219	−0.6189	−0.6158
0.27	−0.6128	−0.6098	−0.6068	−0.6038	−0.6008	−0.5978	−0.5948	−0.5918	−0.5888	−0.5858
0.28	−0.5828	−0.5799	−0.5769	−0.5740	−0.5710	−0.5681	−0.5651	−0.5622	−0.5592	−0.5563
0.29	−0.5534	−0.5505	−0.5476	−0.5446	−0.5417	−0.5388	−0.5359	−0.5330	−0.5302	−0.5273
0.30	−0.5244	−0.5215	−0.5187	−0.5158	−0.5129	−0.5101	−0.5072	−0.5044	−0.5015	−0.4987
0.31	−0.4959	−0.4930	−0.4902	−0.4874	−0.4845	−0.4817	−0.4789	−0.4761	−0.4733	−0.4705
0.32	−0.4677	−0.4649	−0.4621	−0.4593	−0.4565	−0.4538	−0.4510	−0.4482	−0.4454	−0.4427
0.33	−0.4399	−0.4372	−0.4344	−0.4316	−0.4289	−0.4261	−0.4234	−0.4207	−0.4179	−0.4152
0.34	−0.4125	−0.4097	−0.4070	−0.4043	−0.4016	−0.3989	−0.3961	−0.3934	−0.3907	−0.3880
0.35	−0.3853	−0.3826	−0.3799	−0.3772	−0.3745	−0.3719	−0.3692	−0.3665	−0.3638	−0.3611
0.36	−0.3585	−0.3558	−0.3531	−0.3505	−0.3478	−0.3451	−0.3425	−0.3398	−0.3372	−0.3345
0.37	−0.3319	−0.3292	−0.3266	−0.3239	−0.3213	−0.3186	−0.3160	−0.3134	−0.3107	−0.3081
0.38	−0.3055	−0.3029	−0.3002	−0.2976	−0.2950	−0.2924	−0.2898	−0.2871	−0.2845	−0.2819
0.39	−0.2793	−0.2767	−0.2741	−0.2715	−0.2689	−0.2663	−0.2637	−0.2611	−0.2585	−0.2559
0.40	−0.2533	−0.2508	−0.2482	−0.2456	−0.2430	−0.2404	−0.2378	−0.2353	−0.2327	−0.2301
0.41	−0.2275	−0.2250	−0.2224	−0.2198	−0.2173	−0.2147	−0.2121	−0.2096	−0.2070	−0.2045
0.42	−0.2019	−0.1993	−0.1968	−0.1942	−0.1917	−0.1891	−0.1866	−0.1840	−0.1815	−0.1789
0.43	−0.1764	−0.1738	−0.1713	−0.1687	−0.1662	−0.1637	−0.1611	−0.1586	−0.1560	−0.1535
0.44	−0.1510	−0.1484	−0.1459	−0.1434	−0.1408	−0.1383	−0.1358	−0.1332	−0.1307	−0.1282
0.45	−0.1257	−0.1231	−0.1206	−0.1181	−0.1156	−0.1130	−0.1105	−0.1080	−0.1055	−0.1030
0.46	−0.1004	−0.0979	−0.0954	−0.0929	−0.0904	−0.0878	−0.0853	−0.0828	−0.0803	−0.0778
0.47	−0.0753	−0.0728	−0.0702	−0.0677	−0.0652	−0.0627	−0.0602	−0.0577	−0.0552	−0.0527
0.48	−0.0502	−0.0476	−0.0451	−0.0426	−0.0401	−0.0376	−0.0351	−0.0326	−0.0301	−0.0276
0.49	−0.0251	−0.0226	−0.0201	−0.0176	−0.0150	−0.0125	−0.0100	−0.0075	−0.0050	−0.0025

Prob.	.000	.001	.002	.003	.004	.005	.006	.007	.008	.009
0.50	0.0000	0.0025	0.0050	0.0075	0.0100	0.0125	0.0150	0.0176	0.0201	0.0226
0.51	0.0251	0.0276	0.0301	0.0326	0.0351	0.0376	0.0401	0.0426	0.0451	0.0476
0.52	0.0502	0.0527	0.0552	0.0577	0.0602	0.0627	0.0652	0.0677	0.0702	0.0728
0.53	0.0753	0.0778	0.0803	0.0828	0.0853	0.0878	0.0904	0.0929	0.0954	0.0979
0.54	0.1004	0.1030	0.1055	0.1080	0.1105	0.1130	0.1156	0.1181	0.1206	0.1231
0.55	0.1257	0.1282	0.1307	0.1332	0.1358	0.1383	0.1408	0.1434	0.1459	0.1484
0.56	0.1510	0.1535	0.1560	0.1586	0.1611	0.1637	0.1662	0.1687	0.1713	0.1738
0.57	0.1764	0.1789	0.1815	0.1840	0.1866	0.1891	0.1917	0.1942	0.1968	0.1993
0.58	0.2019	0.2045	0.2070	0.2096	0.2121	0.2147	0.2173	0.2198	0.2224	0.2250
0.59	0.2275	0.2301	0.2327	0.2353	0.2378	0.2404	0.2430	0.2456	0.2482	0.2508
0.60	0.2533	0.2559	0.2585	0.2611	0.2637	0.2663	0.2689	0.2715	0.2741	0.2767
0.61	0.2793	0.2819	0.2845	0.2871	0.2898	0.2924	0.2950	0.2976	0.3002	0.3029
0.62	0.3055	0.3081	0.3107	0.3134	0.3160	0.3186	0.3213	0.3239	0.3266	0.3292
0.63	0.3319	0.3345	0.3372	0.3398	0.3425	0.3451	0.3478	0.3505	0.3531	0.3558
0.64	0.3585	0.3611	0.3638	0.3665	0.3692	0.3719	0.3745	0.3772	0.3799	0.3826
0.65	0.3853	0.3880	0.3907	0.3934	0.3961	0.3989	0.4016	0.4043	0.4070	0.4097
0.66	0.4125	0.4152	0.4179	0.4207	0.4234	0.4261	0.4289	0.4316	0.4344	0.4372
0.67	0.4399	0.4427	0.4454	0.4482	0.4510	0.4538	0.4565	0.4593	0.4621	0.4649
0.68	0.4677	0.4705	0.4733	0.4761	0.4789	0.4817	0.4845	0.4874	0.4902	0.4930
0.69	0.4959	0.4987	0.5015	0.5044	0.5072	0.5101	0.5129	0.5158	0.5187	0.5215
0.70	0.5244	0.5273	0.5302	0.5330	0.5359	0.5388	0.5417	0.5446	0.5476	0.5505
0.71	0.5534	0.5563	0.5592	0.5622	0.5651	0.5681	0.5710	0.5740	0.5769	0.5799
0.72	0.5828	0.5858	0.5888	0.5918	0.5948	0.5978	0.6008	0.6038	0.6068	0.6098
0.73	0.6128	0.6158	0.6189	0.6219	0.6250	0.6280	0.6311	0.6341	0.6372	0.6403
0.74	0.6433	0.6464	0.6495	0.6526	0.6557	0.6588	0.6620	0.6651	0.6682	0.6713
0.75	0.6745	0.6776	0.6808	0.6840	0.6871	0.6903	0.6935	0.6967	0.6999	0.7031
0.76	0.7063	0.7095	0.7128	0.7160	0.7192	0.7225	0.7257	0.7290	0.7323	0.7356
0.77	0.7388	0.7421	0.7454	0.7488	0.7521	0.7554	0.7588	0.7621	0.7655	0.7688
0.78	0.7722	0.7756	0.7790	0.7824	0.7858	0.7892	0.7926	0.7961	0.7995	0.8030
0.79	0.8064	0.8099	0.8134	0.8169	0.8204	0.8239	0.8274	0.8310	0.8345	0.8381
0.80	0.8416	0.8452	0.8488	0.8524	0.8560	0.8596	0.8633	0.8669	0.8705	0.8742
0.81	0.8779	0.8816	0.8853	0.8890	0.8927	0.8965	0.9002	0.9040	0.9078	0.9116
0.82	0.9154	0.9192	0.9230	0.9269	0.9307	0.9346	0.9385	0.9424	0.9463	0.9502
0.83	0.9542	0.9581	0.9621	0.9661	0.9701	0.9741	0.9782	0.9822	0.9863	0.9904
0.84	0.9945	0.9986	1.0027	1.0069	1.0110	1.0152	1.0194	1.0237	1.0279	1.0322
0.85	1.0364	1.0407	1.0450	1.0494	1.0537	1.0581	1.0625	1.0669	1.0714	1.0758
0.86	1.0803	1.0848	1.0893	1.0939	1.0985	1.1031	1.1077	1.1123	1.1170	1.1217
0.87	1.1264	1.1311	1.1359	1.1407	1.1455	1.1503	1.1552	1.1601	1.1650	1.1700
0.88	1.1750	1.1800	1.1850	1.1901	1.1952	1.2004	1.2055	1.2107	1.2160	1.2212
0.89	1.2265	1.2319	1.2372	1.2426	1.2481	1.2536	1.2591	1.2646	1.2702	1.2759
0.90	1.2816	1.2873	1.2930	1.2988	1.3047	1.3106	1.3165	1.3225	1.3285	1.3346
0.91	1.3408	1.3469	1.3532	1.3595	1.3658	1.3722	1.3787	1.3852	1.3917	1.3984
0.92	1.4051	1.4118	1.4187	1.4255	1.4325	1.4395	1.4466	1.4538	1.4611	1.4684
0.93	1.4758	1.4833	1.4909	1.4985	1.5063	1.5141	1.5220	1.5301	1.5382	1.5464
0.94	1.5548	1.5632	1.5718	1.5805	1.5893	1.5982	1.6072	1.6164	1.6258	1.6352
0.95	1.6449	1.6546	1.6646	1.6747	1.6849	1.6954	1.7060	1.7169	1.7279	1.7392
0.96	1.7507	1.7624	1.7744	1.7866	1.7991	1.8119	1.8250	1.8384	1.8522	1.8663
0.97	1.8808	1.8957	1.9110	1.9268	1.9431	1.9600	1.9774	1.9954	2.0141	2.0335
0.98	2.0537	2.0749	2.0969	2.1201	2.1444	2.1701	2.1973	2.2262	2.2571	2.2904
0.99	2.3263	2.3656	2.4089	2.4573	2.5121	2.5758	2.6521	2.7478	2.8782	3.0902

B Literature

AGRESTI, A.[1990] Categorical Data Analysis. John Wiley & Sons

AGRESTI, A.[2002] Categorical Data Analysis. John Wiley & Sons 2nd ed.

AGRESTI, A.[2007] An Introduction To Categorical Data Analysis. John Wiley & Sons 2nd ed.

AMRHEIN, V., GREENLAND, S., MCSHANE, B.[2019] Retire Statistical Significance. Nature, Vol. 567, p. 305 - 307, 21 March 2019

ANDERSON, D.R., SWEENEY, D.J., WILLIAMS, TH.A., FREEMAN, J., SHOE-SMITH, E. [2017] Statistics for Business and Economics. Thompson Learning, Thomson London 4th ed.

ARRENBERG, J.[2020] Wirtschaftsstatistik für Bachelor. UTB UVK Lucius 4th ed.

ARRENBERG, J., KOWALSKI, S.[2007] Studie: Lernen Frauen und Männer unterschiedlich? Kompetenzzentrum Technik- Diversity - Chancengleichheit, Bielefeld

BERENSON, M.L., LEVINE, D.M., KREHBIEL, T.C. [2015] Basic Business Statistics. Pearson 13th ed.

BRADLEY, T. [2007] Essential Statistics For Economics, Business and Management. John Wiley& Sons

BROSIUS, F.[2018] SPSS. MITP-Verlag, Heidelberg 8. ed.

BÜHL, A.[2018] SPSS, Einführung in die moderne Datenanalyse. PEARSON Studium, 16. ed.

BÜNING, H., THADEWALD, TH.[2007] Jarque-bera Test and Its Competitors for Testing Normality: A Power Comparison. Journal of Applied Statistics 2007, Volumne 34, issue 1, pages 87-105

CHO, DONG W., IM, KYUNG SO[2002] A Test of Normality Using Geary's Skewness and Kurtosis Statistics. www.bus.ucf.edu/documents/-economics/workingpapers/2002-32.pdf

DANIEL, W. W.[2004] Biostatistics. John Wiley & Sons 7th ed.

FERGUSON, TH. S.[1967] Mathematical Statistics. Academic Press

GIBBONS, J.[2010] Nonparametric Statistical Inference. Marcel Dekker-Verlag, New York, 5th ed.

HANDL, A.[2002] Multivariate Analysemethoden. Springer-Verlag 1. ed.

HANDL, A., KUHLENKASPER, T.[2017] Multivariate Analysemethoden. Theorie und Praxis mit R, Springer-Spektrum 3. ed.

SCHLITTGEN, R.[2012] Einführung in die Statistik, Analyse und Modellierung von Daten. Oldenbourg-Verlag, München 12. ed.

SCHLITTGEN, R.[2009] Multivariate Statistik. Oldenbourg-Verlag 1. ed.

SHAPIRO, S. S., WILK, M. B.[1965] An Analysis of Variance Test for Normality (Complete Samples). Biometrika, Vol. 52, No. 3/4. (Dec., 1965), pp. 591-611 (vgl. http://sci2s.ugr.es/keel/pdf/ algorithm/articulo /shapiro1965.pdf)

C Online Course Materials

youtube Tutorial Introduction to SPSS
https://www.youtube.com/watch?v=SL2bZXfWQls

D Index

leverage values, 107
Likelihood Ratio Test, 136
Lilliefors Test, 57
Log-Likelihood-Chi-Square-Test, 145

mid rank, 28
multicollinearity, 112
multivariate, 3

nominal, 2

ordinal, 2, 20
outlier, 122, 160, 163

P-P plot, 64
p-value, 8
Pearson Chi-Square Test, 14
    layer, 18
    rule of thumb, 17
Phi-Coefficient, 33
proximity matrix, 198

QQ-Plot, 66

rank correlation
    biserial, 31
    of Spearman, 29
regression coefficient, 97
regression line, 102
reliable, 103, 104, 115
residual, 104
    residuum plot, 115
right steep, 59

sample deviation, 104
sample variance, 76
scale, 2
scree plot, 190

Shapiro-Wilk Test, 57
significance, 8
single-linkage-solution, 199
skewness, 58
Spearman's rho, 29
standard deviation, 7
statistical test, 8
stochastically independent, 11, 116

t-Test, 82, 170
    one-sided, 88
test in regression model
    t-Test, 119, 120
    F-Test, 121
tie, 28, 35, 66

uncorrelated, 116
univariate, 3

value label, 25
variance inflation factor, 112

Wald test, 137, 146
Welch Test, 84, 85